# Lecture Notes in Statistics

Edited by P. Bickel, P.J. Diggle, S.E. Fienberg, U. C
I. Olkin, S. Zeger

Paul Doukhan • Gabriel Lang •
Donatas Surgailis • Gilles Teyssière
Editors

# Dependence in Probability and Statistics

 Springer

*Editors*
Prof. Paul Doukhan
UFR Sciences-Techniques
avenue Adolphe-Chauvin 2
95302 Pontoise
France
doukhan@u-cergy.fr

Prof. Dr. Gabriel Lang
INRA AgroParisTech
UMR MIA 518
75005 Paris
France
gabriel.lang@engref.agroparistech.fr

Prof. Dr. Donatas Surgailis
Stochastic Processes Department
Akademijos str. 4
08412 Vilnius
Lithuania
sdonatas@ktl.mii.lt

Dr. Gilles Teyssière
CREATES
School of Economics
Aarhus University
Bartholins Allé 10
8000 Aarhus C
Denmark
stats@gillesteyssiere.net

ISSN 0930-0325
ISBN 978-3-642-14103-4      e-ISBN 978-3-642-14104-1
DOI 10.1007/978-3-642-14104-1
Springer Heidelberg Dordrecht London New York

Library of Congress Control Number: 2010931866

*Cover design*: SPi Publisher Services

Printed on acid-free paper

Springer is part of Springer Science+Business Media (www.springer.com)

# Preface

This volume contains several contributions on the general theme of dependence for several classes of stochastic processes, and its implications on asymptotic properties of various statistics and on statistical inference issues in statistics and econometrics.

The chapter by Berkes, Horváth and Schauer is a survey on their recent results on bootstrap and permutation statistics when the negligibility condition of classical central limit theory is not satisfied. These results are of interest for describing the asymptotic properties of bootstrap and permutation statistics in case of infinite variances, and for applications to statistical inference, e.g., the change-point problem.

The paper by Stoev reviews some recent results by the author on ergodicity of max-stable processes. Max-stable processes play a central role in the modeling of extreme value phenomena and appear as limits of component-wise maxima. At the present time, a rather complete and interesting picture of the dependence structure of max-stable processes has emerged, involving spectral functions, extremal stochastic integrals, mixed moving maxima, and other analytic and probabilistic tools. For statistical applications, the problem of ergodicity or non-ergodicity is of primary importance.

The main statistical problem for long memory processes is estimation of the long memory parameter. This problem has been extensively studied in the literature and optimal convergence rates of certain popular estimators of the memory parameter were established under the assumption that the spectral density is second order regularly varying at zero. The paper by Soulier discusses the situation when the spectral density does not satisfy the second order regularity condition, and obtains a lower bound on the convergence rate of estimator. It turns out that in such a case, the rates of convergence can be extremely slow.

In statistical inference for time series and random fields, the important question is to prove the asymptotic normality of certain quadratic forms written as integrals of the periodogram with certain kernels. By the well-known cumulant criterion, this problem can be reduced to estimation of cumulants of the quadratic form; however, such cumulants have a complicated form of Fejér graph integrals and are generally difficult to analyze. The paper by Avram, Leonenko and Sakhno reviews and extends their earlier asymptotic results on asymptotic behavior of Fejér graph integrals and applies them to prove asymptotic normality of tapered estimators.

Vedel, Wendt, Abry and Jaffard consider two classes of multifractal processes, i.e., infinitely divisible motions and fractional Brownian motions in multifractal time, and three different types of multiresolution quantities: increment, wavelet and Leader coefficients, which are commonly involved in the analysis and modeling of scale invariance in applications. They study, both analytically and by means of numerical simulations, the impact of varying the number of vanishing moments of the mother wavelet on the decay rate of the higher order correlation functions of these multiresolution coefficients. They found that increasing the number of vanishing moments of the mother wavelet significantly fastens the decay of the correlation functions, but does not induce any faster decay for the higher order correlation functions.

It is well-known that real world signals often display erratic oscillations (erratic Hölder exponents) which are not characteristic for "classical" stochastic processes such as diffusions or Gaussian processes, and which have been termed multifractional behavior. An interesting and challenging task for probabilists is to construct new classes of random multifractals and to rigorously describe multifractional properties. The paper by Ahn, Leonenko and Shieh deals with the multifractal products of the processes being the exponents of Ornstein- Uhlenbeck processes driven by Lévy motion. The paper reviews some known results and studies some questions related with tempered stable and normal tempered stable distributions.

The chapter by Wu and Mielniczuk revisits the concept of statistical dependence, viewed as the state of variables being influenced by others. With this understanding of dependence, they introduce new dependence measures which are easy to work with and are useful for developing an asymptotic theory for complicated stochastic systems. They also explore relations of the introduced dependence concept with nonlinear system theory, experimental design, information theory and risk management.

Haye and Farrell's paper develops the limit theory for robust regression estimation when inputs are linear long range dependent processes; limit theory for functions of such models comes from the elegant martingale decomposition of Ho and Hsing (1995, 1996). The difficult point is to establish an uniform inequality affordable following Wu's ideas. As usual, rates depend on the Hurst exponent. Robust regression extends the formerly studied case of least squares error minimization and should be very helpful for dealing with real data.

The same regression models are considered by Schmitz and Steinebach to get an automatic monitoring of changes in linear regression models; CUSUM procedures inherited from Horváth et al. (2004) allow to provide a time segmentation on which the model is a simple regression with dependent inputs. Extremities of the considered intervals are the change-point epochs and almost sure validations of the method are proved; the method is illustrated in different dependence frameworks: the econometrician's Near Epoch Dependence, M-dependence and strong mixing AR. The most interesting feature is that distributional convergences allow to derive asymptotic confidence sets.

Nonstationarity is a deep difficulty when handling with really observed times series. Kouamo, Moulines and Roueff test for homogeneity of times series through a multiscale idea; it is well known that wavelets are associated to difference operators extending on the discrete derivative. The considered retrospective method makes changes into the spectral density become changes of the variance of the wavelet coefficients. Hence a CUSUM procedure is worked out to achieve the authors program for a Gaussian case.

We thank the authors for their contributions, as well as the anonymous referees for their great help.


Paris, Vilnius, Aarhus,                                                                    *Paul Doukhan*
December 2009                                                                              *Gabriel Lang*
                                                                                           *Donatas Surgailis*
                                                                                           *Gilles Teyssière*

# Contents

# List of Contributors

Patrice Abry
ENS Lyon, CNRS UMR 5672, 46 allée d'Italie, 69364 Lyon cedex, e-mail:
patrice.abry@ens-lyon.fr

Vo V. Anh
School of Mathematical Sciences, Queensland University of Technology GPO Box
2434, Brisbane QLD 4001, Australia, e-mail: v.anh@qut.edu.au

Florin Avram
Département de Mathématiques, Université de Pau et des Pays de l'Adour, Avenue
de l'Université - BP 1155 64013 Pau Cedex France e-mail: Florin.Avram@univ-
Pau.fr

István Berkes
Institute of Statistics, Graz University of Technology, Austria e-mail:
berkes@tugraz.at

Patrick Farrell
Carleton University, 1125 Colonel By Dr. Ottawa, Ontario, Canada, K1S 5B6,
e-mail: pfarrell@math.carleton.ca

Lajos Horváth
Department of Mathematics, University of Utah, Salt Lake City, USA e-mail:
horvath@math.utah.edu

Stéphane Jaffard
Université Paris Est, CNRS, 61, avenue du Général de Gaulle, 94010 Créteil,
e-mail: jaffard@univ-paris12.fr

Olaf Kouamo
ENSP, LIMSS, BP : 8390 Yaoundé, e-mail: olaf.kouamo@telecom-paristech.fr

Nikolai N. Leonenko
Cardiff School of Mathematics Cardiff University Senghennydd Road, Cardiff

CF24 4YH, UK, e-mail: LeonenkoN@Cardiff.ac.uk

Jan Mielniczuk
Polish Academy of Sciences Warsaw University of Technology, Warsaw, Poland
e-mail: miel@ipipan.waw.pl

Eric Moulines,
Institut Télécom, Télécom ParisTech, CNRS UMR 5181, Paris, e-mail:
eric.moulines@telecom-paristech.fr

Mohamedou Ould-Haye
Carleton University, 1125 Colonel By Dr. Ottawa, Ontario, Canada, K1S 5B6,
e-mail: ouldhaye@math.carleton.ca

Francois Roueff
Institut Télécom, Télécom ParisTech, CNRS UMR 5181, Paris, e-mail:
francois.roueff@telecom-paristech.fr

Ludmila Sakhno
Department of Probability Theory and Mathematical Statistics, Kyiv National Taras
Shevchenko University, Ukraine e-mail: lms@univ.kiev.ua

Johannes Schauer
Institute of Statistics, Graz University of Technology, Austria e-mail: jo-
hannes.schauer@tugraz.at

Alexander Schmitz
Mathematisches Institut, Universität zu Köln, Weyertal 86-90, 50931 Köln,
Germany. e-mail: schmitza@math.uni-koeln.de

Narn-Rueih Shieh
Department of Mathematics National Taiwan University Taipei 10617, Taiwan,
e-mail: shiehnr@math.ntu.edu.tw

Philippe Soulier
Université Paris Ouest-Nanterre, 200 avenue de la République, 92000 Nanterre
cedex, France, e-mail: philippe.soulier@u-paris10.fr

Josef G. Steinebach
Mathematisches Institut, Universität zu Köln, Weyertal 86-90, 50931 Köln,
Germany. e-mail: jost@math.uni-koeln.de

Stilian A. Stoev
Department of Statistics, The University of Michigan, 439 W. Hall, 1085 S.
University, Ann Arbor, MI 48109–1107, e-mail: sstoev@umich.edu

Béatrice Vedel
Université de Bretagne Sud, Université Européenne de Bretagne, Campus de
Tohannic, BP 573, 56017 Vannes, e-mail: beatrice.vedel@univ-ubs.fr

Herwig Wendt
ENS Lyon, CNRS UMR 5672, 46 allée d'Italie, 69364 Lyon cedex, e-mail:

herwig.wendt@ens-lyon.fr

Wei Biao Wu
University of Chicago, Department of Statistics, The University of Chicago,
Chicago, IL 60637, USA, e-mail: wbwu@galton.uchicago.edu

# Permutation and bootstrap statistics under infinite variance

István Berkes, Lajos Horváth, and Johannes Schauer

**Abstract** Selection from a finite population is used in several procedures in statistics, among others in bootstrap and permutation methods. In this paper we give a survey of some recent results for selection in "nonstandard" situations, i.e. in cases when the negligibility condition of classical central limit theory is not satisfied. These results enable us to describe the asymptotic properties of bootstrap and permutation statistics in case of infinite variances, when the limiting processes contain random coefficients. We will also show that random limit distributions can be avoided by a suitable trimming of the sample, making bootstrap and permutation methods applicable for statistical inference under infinite variances.

## 1 Introduction

Selection from a finite population is a basic procedure in statistics and large sample properties of many classical tests and estimators are closely connected with the asymptotic behavior of sampling variables. Typical examples are bootstrap and permutation statistics, both of which assume a sample $X_1, X_2, \ldots, X_n$ of i.i.d. random variables with distribution function $F$ and then drawing, with or without replacement, $m = m(n)$ elements from the finite set $\{X_1, \ldots, X_n\}$. The usefulness of this procedure is due to the fact that the asymptotic properties of many important func-

István Berkes
Institute of Statistics, Graz University of Technology, Austria
e-mail: berkes@tugraz.at

Lajos Horváth
Department of Mathematics, University of Utah, Salt Lake City, USA
e-mail: horvath@math.utah.edu

Johannes Schauer
Institute of Statistics, Graz University of Technology, Austria
e-mail: johannes.schauer@tugraz.at

P. Doukhan et al. (eds.), *Dependence in Probability and Statistics*,
Lecture Notes in Statistics 200, DOI 10.1007/978-3-642-14104-1_1,
© Springer-Verlag Berlin Heidelberg 2010

tionals of the random variables $X_1^{(n)}, \ldots, X_m^{(n)}$ obtained by resampling are similar to those of the functionals of the original sample $X_1, \ldots, X_n$. There is an extensive literature of bootstrap and permutation statistics in case of populations with finite variance; on the other hand, very little is known in the case of infinite variances. Athreya [2] showed, in the case when the underlying distribution is a stable distribution with parameter $0 < \alpha < 2$, that the normalized partial sums of bootstrap statistics converge weakly to a random limit distribution, i.e. to a distribution function containing random coefficients. Recently, Aue, Berkes and Horváth [3] extended this result to permutation statistics. Note that the elements of a permuted sample are, in contrast to bootstrap, dependent, leading to a different limit distribution.

The purpose of the present paper is to give a survey of the asymptotics of permutation and bootstrap statistics in case of infinite variances, together with applications to statistical inference, e.g. for change point problems. In Section 2 we will show that resampling from an i.i.d. sample with infinite variance requires studying the limiting behavior of a triangular array of random variables violating the classical uniform asymptotic negligibility condition of central limit theory. Starting with the 1960's, classical central limit theory has been extended to cover such situations (see e.g. Bergström [4]). In this case the limit distribution is generally not Gaussian and depends on the non-negligible elements of the array. In the case of permutation statistics, the row elements of our triangular array are dependent random variables, making the situation considerably more complicated. Theorems 2.3 and 2.4 in Section 2 describe the new situation. As we will show in Section 3, the probabilistically interesting, but statistically undesirable phenomenon of random limit distributions can be avoided by trimmimg the sample, enabling one to extend a number of statistical procedures for observations with infinite variances.

Our paper is a survey of some recent results of the authors; the proofs will be given in our forthcoming papers [6] and [7].

## 2 Some general sampling theorems

For each $n \in \mathbb{N}$ let

$$x_{1,n} \leq x_{2,n} \leq \ldots \leq x_{n,n}$$

be a sequence of real numbers and denote by $X_1^{(n)}, X_2^{(n)}, \ldots, X_m^{(n)}$ the random variables obtained by drawing, with or without replacement, $m = m(n)$ elements from the set $\{x_{1,n}, \ldots, x_{n,n}\}$. Define the partial sum process

$$Z_{n,m}(t) = \sum_{j=1}^{\lfloor mt \rfloor} X_j^{(n)} \qquad \text{for } 0 \leq t \leq 1, \tag{1}$$

where $\lfloor \cdot \rfloor$ denotes integral part. Let $\xrightarrow{\mathscr{D}[0,1]}$ denote convergence in the space $\mathscr{D}[0,1]$ of càdlàg functions equipped with the Skorokhod $J_1$-topology. The following two results are well known.

**Theorem 2.1.** *Let*

$$\sum_{j=1}^{n} x_{j,n} = 0, \qquad \sum_{j=1}^{n} x_{j,n}^2 = 1 \qquad (2)$$

*and*

$$\max_{1 \le j \le n} |x_{j,n}| \longrightarrow 0 \qquad (3)$$

*and draw $m = m(n)$ elements from the set $\{x_{1,n}, \ldots, x_{n,n}\}$ with replacement, where*

$$m/n \to c \qquad \text{for some } c > 0. \qquad (4)$$

*Then*

$$Z_{n,m}(t) \xrightarrow{\mathscr{D}[0,1]} W(ct) \qquad \text{for } n \to \infty,$$

*where $\{W(t), 0 \le t \le 1\}$ is a Wiener process.*

**Theorem 2.2.** *Assume (2) and (3) and draw $m = m(n)$ elements from the set $\{x_{1,n}, \ldots, x_{n,n}\}$ without replacement, where $m \le n$ and (4) holds. Then*

$$Z_{n,m}(t) \xrightarrow{\mathscr{D}[0,1]} B(ct) \qquad \text{for } n \to \infty,$$

*where $\{B(t), 0 \le t \le 1\}$ is a Brownian bridge.*

In the case of Theorem 2.1 the random variables $X_1^{(n)}, \ldots, X_m^{(n)}$ are i.i.d. with mean 0 and variance $1/n$ and they satisfy the Lindeberg condition

$$\lim_{n \to \infty} \sum_{j=1}^{m} E[(X_j^{(n)})^2 I\{|X_j^{(n)}| \ge \varepsilon\}] = 0 \qquad \text{for any } \varepsilon > 0, \qquad (5)$$

since the sum on the left hand side is 0 for $n \ge n_0(\varepsilon)$ by the uniform asymptotic negligibility condition (3). Thus Theorem 2.1 is an immediate consequence of the classical functional central limit theorem for sums of independent random variables (see e.g. Skorokhod [16]). Theorem 2.2, due to Rosén [15], describes a different situation: if we sample without replacement, the r.v.'s $X_1^{(n)}, \ldots, X_m^{(n)}$ are dependent and the partial sum process $Z_{n,m}(t)$ converges weakly to a process with dependent (actually negatively correlated) increments.

Typical applications of Theorem 2.1 and Theorem 2.2 include bootstrap and permutation statistics. Let $X_1, X_2, \ldots$ be i.i.d. random variables with distribution function $F$ with mean 0 and variance 1. Let $\{X_1^{(n)}, \ldots, X_m^{(n)}\}$ be the bootstrap sample obtained by drawing $m = m(n)$ elements from the set $\{X_1, \ldots, X_n\}$ with replacement. Clearly, $X_1^{(n)}, \ldots, X_m^{(n)}$ are independent random variables with common distribution $F_n(t) = (1/n) \sum_{i=1}^{n} I\{X_i \le t\}$, the empirical distribution function of the sample $X_1, \ldots, X_n$. Define

$$\overline{X}_n = \frac{1}{n}\sum_{k=1}^{n} X_k \quad \text{and} \quad \sigma_n^2 = \frac{1}{n}\sum_{k=1}^{n}(X_k - \overline{X}_n)^2$$

and apply Theorem 2.1 for the random finite set

$$\left\{ \frac{X_1 - \overline{X}_n}{\sigma_n\sqrt{n}}, \ldots, \frac{X_n - \overline{X}_n}{\sigma_n\sqrt{n}} \right\}, \tag{6}$$

where the selection process is independent of the sequence $X_1, X_2, \ldots$. It is easily checked that the conditions of Theorem 2.1 are satisfied and it follows that if (4) holds then conditionally on $\mathbf{X} = (X_1, X_2, \ldots)$, for almost all paths $(X_1, X_2, \ldots)$,

$$P_{\mathbf{X}}\left\{ \frac{1}{\sigma_n\sqrt{n}}\sum_{k=1}^{\lfloor mt \rfloor}(X_k^{(n)} - \overline{X}_n) \xrightarrow{\mathscr{D}[0,1]} W(ct) \right\} = 1.$$

This fundamental limit theorem for the bootstrap is due to Bickel and Freedman [8]. On the other hand, drawing $n$ elements from the set $\{X_1, \ldots, X_n\}$ without replacement, we get a random permutation of $X_1, \ldots, X_n$ which we denote by $X_{\pi(1)}, \ldots, X_{\pi(n)}$. Again we assume that the selection process is independent of $X_1, X_2, \ldots$. It is clear that all $n!$ permutations of $(X_1, X_2, \ldots, X_n)$ are equally likely. Applying now Theorem 2.2 for the set (6), we get that for almost all paths $\mathbf{X} = (X_1, X_2, \ldots)$,

$$P_{\mathbf{X}}\left\{ \frac{1}{\sigma_n\sqrt{n}}\sum_{k=1}^{\lfloor nt \rfloor}(X_{\pi(k)} - \overline{X}_n) \xrightarrow{\mathscr{D}[0,1]} B(t) \right\} = 1,$$

an important fact about permutation statistics.

As the above results show, the uniform asymptotic negligibility condition for the random set (6) is satisfied if $EX_1^2 < \infty$. It is easy to see that the converse is also true. Thus studying bootstrap and permutation statistics under infinite variances requires a model where uniform asymptotic negligibility fails, i.e. the elements of the set $\{x_{1,n}, \ldots, x_{n,n}\}$ are not any more "small". Clearly, in this case the limiting behavior of the partial sums of the selected elements will be quite different. If, for example, the largest element $x_{n,n}$ of the set $\{x_{1,n}, \ldots, x_{n,n}\}$ does not tend to 0 as $n \to \infty$, then the contribution of $x_{n,n}$ in the partial sums of a sample of size $n$ taken from this set clearly will not be negligible and thus the limit distribution of such sums (if exists) will depend on this largest element. A similar effect is well known in classical central limit theory (see e.g. Bergström [4]), but the present situation will exhibit substantial additional difficulties. Without loss of generality we may assume again that

$$\sum_{j=1}^{n} x_{j,n} = 0. \tag{7}$$

Next we will assume

$$x_{j,n} \longrightarrow y_j \quad \text{and} \quad x_{n-j+1,n} \longrightarrow z_j \tag{8}$$

for any fixed $j$ as $n \to \infty$ for some numbers $y_j, z_j, j \in \mathbb{N}$ and that

$$\lim_{K \to \infty} \limsup_{n \to \infty} \sum_{j=K+1}^{n-K} x_{j,n}^2 = 0. \tag{9}$$

Condition (8) is no essential restriction of generality: if we assume only that the sequences $\{x_{j,n}, n \geq 1\}$, $\{x_{n-j+1,n}, n \geq 1\}$ are bounded for any fixed $j$, then by a diagonal argument we can find a subsequence of $n$'s along which (8) holds. Then along this subsequence our theorems will apply and if the limiting numbers $y_j, z_j$ are different along different subsequences, the processes $Z_{n,m}(t)$ will also have different limits along different subsequences. This seems to be rather pathological behavior, but it can happen even in simple i.i.d. situations, see Corollary 2.5 below. The role of condition (9) is to exclude a Wiener or Brownian bridge component in the limiting process, as it occurs in Theorem 2.1.

We formulate now our main sampling theorems. For the proof we refer to Berkes, Horváth and Schauer [6].

**Theorem 2.3.** *Let, for each $n = 1, 2, \ldots$,*

$$x_{1,n} \leq x_{2,n} \leq \ldots \leq x_{n,n} \tag{10}$$

*be a finite set satisfying (7), (8), (9) and*

$$\sum_{j=1}^{\infty} y_j^2 < \infty \qquad \text{and} \qquad \sum_{j=1}^{\infty} z_j^2 < \infty. \tag{11}$$

*Let $X_1^{(n)}, \ldots, X_m^{(n)}$ be the random elements obtained by drawing $m = m(n) \leq n$ elements from the set (10) without replacement, where (4) holds. Then for the processes $Z_{n,m}(t)$ defined by (1) we have*

$$Z_{n,m}(t) \xrightarrow{\mathscr{D}[0,1]} R(ct) \qquad \text{for } n \to \infty,$$

*where*

$$R(t) = \sum_{j=1}^{\infty} y_j(\delta_j(t) - t) + \sum_{j=1}^{\infty} z_j(\delta_j^*(t) - t)$$

*and $\{\delta_j(t), 0 \leq t \leq 1\}$, $\{\delta_j^*(t), 0 \leq t \leq 1\}$, $j = 1, 2, \ldots$ are independent jump processes, each making a single jump from 0 to 1 at a random point uniformly distributed in $(0,1)$.*

**Theorem 2.4.** *Let, for each $n = 1, 2, \ldots$,*

$$x_{1,n} \leq x_{2,n} \leq \ldots \leq x_{n,n} \tag{12}$$

*be a finite set satisfying (7), (8), (9) and (11). Let $X_1^{(n)}, \ldots, X_m^{(n)}$ be the random elements obtained by drawing $m = m(n)$ elements from the set (12) with replacement, where (4) holds. Then for the processes $Z_{n,m}(t)$ defined by (1) we have*

$$Z_{n,m}(t) \xrightarrow{\mathscr{D}[0,1]} R(ct) \qquad for \ n \to \infty$$

*where*

$$R(t) = \sum_{j=1}^{\infty} y_j(\delta_j(t) - t) + \sum_{j=1}^{\infty} z_j(\delta_j^*(t) - t)$$

*and* $\{\delta_j(t), t \geq 0\}$, $\{\delta_j^*(t), t \geq 0\}$, $j = 1, 2, \ldots$ *are independent Poisson processes with parameter 1.*

These results show that in the case when the asymptotic negligibility condition is not satisfied, the limiting process will depend on the values of $y_j$ and $z_j$, $j = 1, 2, \ldots$. Obviously the $y_j$'s and $z_j$'s form non-increasing sequences. The larger the differences between consecutive values are, the more the process $R(t)$ will be different from a Wiener process (or a Brownian bridge, respectively). Observe also that in the case of permutation statistics with $m = n$, the process $R(t)$ satisfies $R(0) = R(1) = 0$ and therefore it gives a non-Gaussian "bridge", having the same covariances (up to a constant) as a Brownian bridge. In the bootstrap case $R(t)$ has the same covariances (again up to a constant) as a scaled Wiener process.

Figure 1 shows the sample paths of the (appropriately normalized) limiting process $R(t)$ for permutations with $-y_j = z_j = c j^{-1/a}$ ($a \in \{0.5, 0.8, 1, 1.5\}$) and of a Brownian bridge $B(t)$. The pictures show the differences between the two limiting processes and that increasing the value of $a$ makes $R(t)$ look closer to a Brownian bridge.

We now turn to applications of Theorems 2.3 and 2.4. The simplest situation with infinite variances is the case of i.i.d. random variables $X_1, X_2, \ldots$ belonging to the domain of attraction of a stable r.v. $\xi_\alpha$ with parameter $\alpha \in (0, 2)$. This means that for some numerical sequences $\{a_n, n \in \mathbb{N}\}$, $\{b_n, n \in \mathbb{N}\}$

$$\frac{S_n - a_n}{b_n} \xrightarrow{d} \xi_\alpha, \tag{13}$$

where $S_n = \sum_{j=1}^{n} X_j$. The necessary and sufficient condition for this is

$$P(X_1 > t) \sim p L(t) t^{-\alpha} \quad and \quad P(X_1 < -t) \sim q L(t) t^{-\alpha} \quad as \ t \to \infty \tag{14}$$

for some numbers $p \geq 0$, $q \geq 0$, $p + q = 1$ and a slowly varying function $L(t)$. Let the ordered statistics of $(X_1, \ldots, X_n)$ be $X_{1,n} \leq X_{2,n} \leq \ldots \leq X_{n,n}$ and apply Theorem 2.3 and Theorem 2.4 to the random set

$$\left\{ \frac{X_{1,n} - \bar{X}_n}{T_n}, \ldots, \frac{X_{n,n} - \bar{X}_n}{T_n} \right\}, \tag{15}$$

where $\bar{X}_n = (1/n) \sum_{j=1}^{n} X_j$ is the sample mean and $T_n = \max_{1 \leq j \leq n} |X_j|$. The normalization $T_n$ is used since the r.v.'s are outside the domain of attraction of a normal random variable. This leads to the following results from Aue, Berkes and Horváth [3] and Berkes, Horváth and Schauer [6].

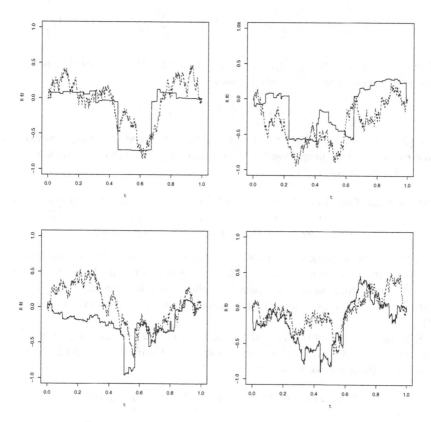

**Fig. 1** Simulations of $R(t)$ (solid) and $B(t)$ (dashed) with $-y_j = z_j = const \cdot j^{-1/a}$ and $a = 0.5, 0.8, 1, 1.5$ (from top left to bottom right)

**Corollary 2.1.** *Let $X_1, X_2, \ldots$ be i.i.d. random variables with partial sums satisfying (13) with some $\{a_n, n \in \mathbb{N}\}$, $\{b_n, n \in \mathbb{N}\}$ and a stable random variable $\xi_\alpha$, $\alpha \in (0,2)$. Furthermore let $X_1^{(n)}, \ldots, X_m^{(n)}$ be the variables obtained by drawing (independently of $X_1, X_2, \ldots$) $m = m(n) \leq n$ times without replacement from the set $\{X_1, \ldots, X_n\}$. Assume that (4) holds and define the functional CUSUM statistic by*

$$Z_{n,m}^*(t) = \frac{1}{T_n} \sum_{j=1}^{\lfloor mt \rfloor} (X_j^{(n)} - \bar{X}_n) \qquad for\ t \in [0,1]. \tag{16}$$

*Then*

$$P_{\mathbf{X}}(Z_{n,m}^*(t) \leq x) \xrightarrow{d} P_{\mathbf{Z}}(R^*(ct) \leq x) \qquad for\ n \to \infty \tag{17}$$

*for any real x, where*

$$R^*(t) = \frac{1}{M} \left( -q^{1/\alpha} \sum_{j=1}^{\infty} \frac{1}{Z_j^{1/\alpha}} (\delta_j(t) - t) + p^{1/\alpha} \sum_{j-1}^{\infty} \frac{1}{(Z_j^*)^{1/\alpha}} (\delta_j^*(t) - t) \right).$$

Here $Z_j = \eta_1 + \ldots + \eta_j$, $Z_j^* = \eta_1^* + \ldots + \eta_j^*$, where $\{\eta_j, \eta_j^*, j \in \mathbb{N}\}$ are independent exponential random variables with parameter 1,

$$M = \max\{(q/Z_1)^{1/\alpha}, (p/Z_1^*)^{1/\alpha}\} \tag{18}$$

and $\{\delta_j(t), t \in [0,1]\}$, $\{\delta_j^*(t), t \in [0,1]\}$, $j \in \mathbb{N}$, are independent jump processes, each making a single jump from 0 to 1 at a random point uniformly distributed in $(0,1)$, also independent of $\{Z_j, Z_j^*, j \in \mathbb{N}\}$.

**Corollary 2.2.** *Corollary 2.1 remains valid if* $X_1^{(n)}, \ldots, X_m^{(n)}$ *are obtained by drawing with replacement from the set* $\{X_1, \ldots, X_n\}$. *In this case* $\{\delta_j(t), t \in [0,1]\}$, $\{\delta_j^*(t), t \in [0,1]\}$, $j \in \mathbb{N}$, *will be independent Poisson processes with parameter 1.*

Note that in both corollaries the right-hand side of (17) is a conditional probability given the random variables $Z_1, Z_1^*, Z_2, Z_2^*, \ldots$. This means that the limit distribution is a random distribution function, possibly defined on a different probability space than $X_1, X_2, \ldots$. In the bootstrap case this phenomenon was first observed by Athreya [2].

It is interesting to note that in the extreme case of $\alpha = 0$, i.e. in case of i.i.d. random variables $X_1, X_2, \ldots$ with slowly varying tails, Corollaries 2.1 and 2.2 remain valid. More precisely, we assume that

$$P(X_1 > t) \sim pL(t) \quad \text{and} \quad P(X_1 < -t) \sim qL(t) \tag{19}$$

for $t \to \infty$ and some non-increasing slowly varying function $L(t)$ that satisfies $\lim_{t \to \infty} L(t) = 0$. Using Theorems 2.3 and 2.4 yields the following results.

**Corollary 2.3.** *Let* $X_1, X_2, \ldots$ *be i.i.d. random variables with slowly varying tails satisfying* (19). *Furthermore let* $X_1^{(n)}, \ldots, X_m^{(n)}$ *be the variables obtained by drawing (independently of* $X_1, X_2, \ldots$*)* $m = m(n) \leq n$ *times without replacement from the set* $\{X_1, \ldots, X_n\}$. *Assume that (4) holds. Define* $Z_{n,m}^*(t)$ *as in (16). Then*

$$P_{\mathbf{X}}(Z_{n,m}^*(t) \leq x) \xrightarrow{d} P_{\mathbf{U}}(R^*(ct) \leq x) \quad \text{for } n \to \infty \tag{20}$$

*for any real x, where*

$$R^*(t) = -I(U > p)(\delta(t) - t) + I(U \leq p)(\delta^*(t) - t).$$

Here U is a uniform random variable on $(0,1)$ and $\{\delta(t), t \in [0,1]\}$, $\{\delta^*(t), t \in [0,1]\}$ are independent jump processes, each making a single jump from 0 to 1 at a random point uniformly distributed in $(0,1)$, independent of U.

**Corollary 2.4.** *Corollary 2.3 remains valid if we sample* $X_1^{(n)}, \ldots, X_m^{(n)}$ *with replacement from the set* $\{X_1, \ldots, X_n\}$ *with* $m = m(n)$ *satisfying (4). Then, however,*

$\{\delta(t), t \in [0,1]\}$ and $\{\delta^*(t), t \in [0,1]\}$ are independent Poisson processes with parameter 1 (independent of $U$).

Our next corollary describes a situation when relation (8) fails, i.e. the sequences $x_{j,n}$ and $x_{n-j+1,n}$ do not converge for fixed $j$. Let $X_1, X_2, \ldots$ be i.i.d. symmetric random variables with the distribution

$$P(X_1 = \pm 2^k) = 2^{-(k+1)} \qquad k = 1,2,\ldots \qquad (21)$$

This is the two-sided version of the St. Petersburg distribution. The distribution function $F(x)$ of $X_1$ satisfies

$$1 - F(x) = 2^{-k} \qquad \text{for } 2^{k-1} \le x < 2^k$$

which shows that $G(x) = x(1 - F(x))$ is logarithmically periodic: if $x$ runs through the interval $[2^k, 2^{k+1})$, then $G(x)$ runs through all values in $[1/2, 1)$ and $G(\log_2 x)$ is periodic with period 1. Thus (14) fails and consequently $F$ does not belong to the domain of attraction of a stable law. The partial sums $S_k = \sum_{k=1}^n X_k$ have a remarkable behavior: for any fixed $1 \le c < 2$, the normed sums $n^{-1}S_n$ converge weakly along the subsequence $n_k = \lfloor c2^k \rfloor$ to an infinitely divisible distribution $F_c$ such that $F_c = F_1^{*c}$ and $F_2 = F_1$. The class $\mathscr{F} = \{F_c, 1 \le c \le 2\}$ can be considered a circle, and in each interval $[2^k, 2^{k+1})$, the distribution of $n^{-1}S_n$ essentially runs around this circle in the sense that $n^{-1}S_n$ is close in distribution to $F_c$ with $c = n/2^k$. This behavior was discovered by S. Csörgő [10], who called this quasiperiodic behavior 'merging'. As the following corollary shows, merging will also take place in the behavior of permutation and bootstrap statistics. For simplicity, we consider the case when we draw $n$ elements from the sample $(X_1, \ldots, X_n)$. Let $\Psi(x)$, $0 < x < \infty$ denote the function which increases linearly from 1/2 to 1 on each interval $(2^j, 2^{j+1}]$, $j = 0, \pm 1, \pm 2, \ldots$.

**Corollary 2.5.** Let $X_1, X_2, \ldots$ be i.i.d. random variables with the distribution (21). Let $X_1^{(n)}, \ldots, X_n^{(n)}$ be the elements obtained by drawing $n$ times with replacement from the set $\{X_1, \ldots, X_n\}$ and let $Z_n^*(t)$ be defined by (16) with $m = n$. Let $1 \le c < 2$. Then for $n_k = \lfloor c2^k \rfloor$ we have

$$P_\mathbf{X}(Z_{n_k}^*(t) \le x) \overset{d}{\longrightarrow} P_\mathbf{Z}(R_c(t) \le x)$$

for any real x, where

$$R_c(t) = \frac{1}{M}\left[ -\sum_{j=1}^\infty \frac{1}{Z_j}\Psi\left(\frac{Z_j}{c}\right)(\delta_j(t) - t) + \sum_{j=1}^\infty \frac{1}{Z_j^*}\Psi\left(\frac{Z_j^*}{c}\right)(\delta_j^*(t) - t) \right]$$

with

$$M = \max\left\{ \frac{\Psi(Z_1/c)}{Z_1}, \frac{\Psi(Z_1^*/c)}{Z_1^*} \right\}.$$

*Here $Z_j = \eta_1 + \ldots + \eta_j$ and $Z_j^* = \eta_1^* + \ldots + \eta_j^*$, where $\{\eta_j, \eta_j^*, j \in \mathbb{N}\}$ are i.i.d. $\exp(1)$ random variables and $\{\delta_j(t), 0 \le t \le 1\}$, $\{\delta_j^*(t), 0 \le t \le 1\}$, $j = 1, 2, \ldots$ are independent jump processes, each making a single jump from $0$ to $1$ at a uniformly distributed point in $(0, 1)$.*

Just like in the case of partial sums, the class $R_c$ of limiting processes is logarithmically periodic, namely $R_{2c} = R_c$ and for a fixed $n$ with $2^k \le n < 2^{k+1}$ the conditional distribution of $Z_n^*(t)$ is close to that of $R_c(t)$ with $c = n/2^k$.

Corollary 2.5 remains valid if we draw $X_1^{(n)}, \ldots, X_n^{(n)}$ without replacement from the set $\{X_1, \ldots, X_n\}$. Then $\delta_j(t)$ and $\delta_j^*(t)$ are independent Poisson processes with parameter 1.

# 3 Application to change point detection

Due to the random limit distributions in Corollaries 2.1 and 2.2, bootstrap and permutation methods cannot be directly used in statistical inference when the observations do not have finite variances. Let $X_1, X_2, \ldots$ be i.i.d. random variables belonging to the domain of normal attraction of a stable r.v. $\xi_\alpha$ with parameter $\alpha \in (0, 2)$. Let $S_n = X_1 + \ldots + X_n$ and let $X_{1,n} \le X_{2,n} \le \ldots \le X_{n,n}$ be the order statistics of the sample $(X_1, \ldots, X_n)$. It is well known (see e.g. Darling [11]) that for any fixed $j$ the ratios $X_{j,n}/S_n$, $X_{n-j,n}/S_n$ have nondegenerate limit distributions as $n \to \infty$, which means that the contribution of the extreme elements in the normed sum $n^{-1/\alpha}(X_1 + \ldots + X_n)$ is not negligible. As Corollaries 2.1 and 2.2 show, both in the permutation and bootstrap case we have

$$P_{\mathbf{X}}(Z_{n,m}^*(t) \le x) \xrightarrow{d} P_{\mathbf{Z}}(R^*(ct) \le x) \qquad \text{as } n \to \infty$$

for any real $x$, provided $m/n \to c > 0$, where

$$R^*(t) = \frac{1}{M}\left( -q^{1/\alpha} \sum_{j=1}^{\infty} \frac{1}{Z_j^{1/\alpha}}(\delta_j(t) - t) + p^{1/\alpha} \sum_{j=1}^{\infty} \frac{1}{(Z_j^*)^{1/\alpha}}(\delta_j^*(t) - t) \right).$$

Here the terms $Z_j^{-1/\alpha}(\delta_j(t) - t)$ and $(Z_j^*)^{-1/\alpha}(\delta_j^*(t) - t)$ are due to the extremal terms $X_{j,n}$ and $X_{n-j,n}$ in the sum $S_n$ and thus to get a limit distribution not containing random coefficients we have to eliminate the effect of the extremal elements. Natural ideas are trimming the sample $(X_1, \ldots, X_n)$ before resampling, or to choose the sample size in resampling as $o(n)$, reducing the chance of the largest elements of the sample to get into the new sample. In this section we will show that after a suitable trimming, the limit distribution of the partial sums of the resampled elements will be normal, and thus bootstrap and permutation methods will work under infinite variances. We will illustrate this with an application to change point detection.

Consider the location model

$$X_j = \mu + \delta I(j > K) + e_j \qquad \text{for } j = 1, \ldots, n, \tag{22}$$

where $1 \le K \le n$, $\mu$ and $\delta = \delta_n \ne 0$ are unknown parameters. We assume that $|\delta| \le D$ with some $D > 0$ and that $e_1, \ldots, e_n$ are i.i.d. random variables with

$$Ee_1 = 0, \quad Ee_1^2 = \sigma^2 > 0 \quad \text{and } E|e_1|^\nu < \infty \text{ with some } \nu > 2. \tag{23}$$

We want to test the hypothesis $H_0 : K \ge n$ against $H_1 : K < n$. Common test statistics for this setting are the CUSUM statistics defined by

$$T_n = \max_{1 \le k \le n} \frac{1}{n^{1/2} \widehat{\sigma}_n} \left| \sum_{j=1}^k (X_j - \bar{X}_n) \right|$$

and

$$T_{n,1} = \max_{1 \le k \le n} \sqrt{\frac{n}{k(n-k)}} \frac{1}{\widehat{\sigma}_n} \left| \sum_{j=1}^k (X_j - \bar{X}_n) \right|,$$

where

$$\bar{X}_n = \frac{1}{n} \sum_{j=1}^n X_j \quad \text{and} \quad \widehat{\sigma}_n^2 = \frac{1}{n} \sum_{j=1}^n (X_j - \bar{X}_n)^2.$$

The limit distributions of both statistics are known:

$$\lim_{n \to \infty} P_{H_0}(T_n \le x) = P\left( \sup_{0 \le t \le 1} |B(t)| \le x \right),$$

and

$$\lim_{n \to \infty} P_{H_0}\left( a(\log n) T_{n,1} \le x + b(\log n) \right) = \exp(-2\exp(-x)),$$

where $\{B(t), t \in [0,1]\}$ is a Brownian bridge,

$$a(x) = \sqrt{2\log x} \qquad \text{and} \qquad b(x) = 2\log x + \frac{1}{2} \log\log x - \frac{1}{2} \log \pi.$$

(see Csörgő and Horváth [9] and Antoch and Hušková [1]) and one can determine critical values based on these limit distributions. On the other hand, under the change point alternative if $n\delta_n^2 \to \infty$, then $T_n \to \infty$ in probability and if $n\delta_n^2/\log\log n \to \infty$, then $T_n/\sqrt{\log\log n} \to \infty$ in probability. However, the convergence to the limit distribution is rather slow under the null hypothesis and thus the obtained critical values will work only for large sample sizes and lead to conservative tests otherwise. Antoch and Hušková [1] proposed the use of permutation statistics to get correct critical values for small and moderate size samples. Consider a sample $\mathbf{X} = (X_1, \ldots, X_n)$, let $\pi = (\pi(1), \ldots, \pi(n))$ be a random permutation of $(1, \ldots, n)$, independent of the sample $(X_1, \ldots, X_n)$, and let $\mathbf{X}_\pi = (X_{\pi(1)}, \ldots, X_{\pi(n)})$ be the permuted sample. Let $T_{n,1}^* = T_{n,1}(\mathbf{X}_\pi)$, where $\mathbf{X}$ is considered fixed, and the randomness is in $\pi$. Antoch and Hušková [1] showed the following theorem:

**Theorem 3.1.** *If conditions* (22) *and* (23) *are satisfied and* $|\delta_n| \leq D$ *for some* $D > 0$, *then for all* $x$

$$\lim_{n \to \infty} P_{\mathbf{X}}(T_n^* \leq x) = P\left(\sup_{0 \leq t \leq 1} |B(t)| \leq x\right)$$

*and*

$$\lim_{n \to \infty} P_{\mathbf{X}}\left(a(\log n)T_{n,1}^* \leq x + b(\log n)\right) = \exp(-2\exp(-x)),$$

*for almost all realizations of* $\mathbf{X}$ *as* $n \to \infty$.

Note that Theorem 3.1 is valid under the null as well as under the alternative hypotheses. This shows that the critical values for the test based on $T_{n,1}$ can be replaced by the sample quantiles of its permutation version based on $T_{n,1}^*$, a procedure which is numerically quite convenient. For a given sample $(X_1, \ldots, X_n)$ we generate a large number of random permutations which leads to the empirical distributions of $T_n(\mathbf{X}_\pi)$ and $T_{n,1}(\mathbf{X}_\pi)$ and hence to the desired critical values. As the simulations in [1] show, these critical values are much more satisfactory than the critical values based on the limit distribution. As Hušková [14] pointed out, approximations using bootstrap versions of the test statistics would also work well.

Using $N$ independent permutations, that is $N$ values of $T_n^*$, denoted by $T_n^*(j)$, $j = 1, 2, \ldots, N$, let

$$H_{n,N}(x) = \frac{1}{N} \sum_{j=1}^{N} \{T_n^*(j) \leq x\},$$

be the empirical distribution function that can be used to approximate

$$H_n(x) = P_{H_0}(T_n \leq x).$$

Define $H_{n,N,1}$ and $H_{n,1}$ as the analogues to $H_{n,N}$ and $H_n$ where $T_{n,1}$ replaces $T_n$. Berkes, Horváth, Hušková and Steinebach [5] showed that if conditions (22) and (23) are satisfied, then we have

$$|H_{n,N}(x) - H_n(x)| = o_{P_{\mathbf{X}}}(1) \qquad \text{as } \min(n, N) \to \infty$$

for almost all realizations of $\mathbf{X}$. They also studied the rate of convergence and they proved

$$|H_{n,N}(x) - H_n(x)| = \mathscr{O}_{P_{\mathbf{X}}}\left(N^{-1/2} + n^{-(v-2)/(6v)}\right)$$

and

$$|H_{n,N,1}(x) - H_{n,1}(x)| = o_{P_{\mathbf{X}}}\left(N^{-1/2} + (\log\log n)^{-v/2}\right)$$

for almost all realizations of $\mathbf{X}$.

The previous results show that permutation and bootstrap statistics provide an effective way to detect a change of location in an i.i.d. sequence $(X_n)$. Note, however, that for the validity of the limit distribution results above we need, by (23), the existence of $v > 2$ moments of the underlying variables. As we will see below, using a suitable trimming of the sample $(X_1, \ldots, X_n)$, the limiting processes in Corollaries 2.1 and 2.2 become Brownian motion, resp. Brownian bridge, and then the bootstrapped or permuted version of the CUSUM statistics will work without assuming the existence of second moments.

Fix a sequence $\omega_n \to \infty$ of integers with $\omega_n/n \to 0$. Put $m = n - 2\omega_n$ and let $(Y_1,\ldots,Y_m)$ denote the part of the original sample $(X_1,\ldots,X_n)$ obtained by removing the $\omega_n$ smallest and largest elements from the set. Let $Y_{1,m} \le \ldots \le Y_{m,m}$ be the ordered sample of $(Y_1,\ldots,Y_m)$. Draw $m$ elements $Y_1^{(m)},\ldots,Y_m^{(m)}$ from the set $\{Y_1,\ldots,Y_m\}$ with or without replacement. Let $\varepsilon_j^{(m)}(t)$ count how many times $Y_{j,m}$ has been chosen among the first $\lfloor mt \rfloor$ sampled elements:

$$\varepsilon_j^{(m)}(t) = k \text{ if } Y_{j,m} \text{ has been chosen } k \text{ times among the first } \lfloor mt \rfloor \text{ elements,}$$

for $j = 1,\ldots,m$. Clearly, in the case of selection without replacement $k$ can only take the values 0, 1 while $k \in \{0,1,\ldots,\lfloor mt \rfloor\}$ when drawing with replacement. Letting $\bar{Y}_m = (1/m)\sum_{j=1}^m Y_j = (1/m)\sum_{j=1}^m Y_{j,m}$, we have

$$\widehat{Z}_m(t) := \sum_{j=1}^{\lfloor mt \rfloor} (Y_j^{(m)} - \bar{Y}_m) = \sum_{j=1}^m (Y_{j,m} - \bar{Y}_m)\varepsilon_j^{(m)}(t) = \sum_{j=1}^m (Y_{j,m} - \bar{Y}_m)\bar{\varepsilon}_j^{(m)}(t),$$

where $\bar{\varepsilon}_j^{(m)}(t) = \varepsilon_j^{(m)}(t) - E\varepsilon_j^{(m)}(t)$ is the centered version of $\varepsilon_j^{(m)}(t)$.

**Theorem 3.2.** *In the case of selection without replacement, there exist independent, identically distributed indicator variables $\delta_j^{(m)}(t)$, $j = 1,\ldots,m$ with $P(\delta_j^{(m)}(t) = 1) = \lfloor mt \rfloor/m$ such that for any $0 < t < 1$*

$$P\left( \varepsilon_j^{(m)}(t) \ne \delta_j^{(m)}(t) \quad \text{for some } 1 \le j \le m^{1/3} \text{ or } m - m^{1/3} < j \le m \right)$$

$$\le Ct^{-1}m^{-1/6}, \tag{24}$$

*where C is an absolute constant. Moreover, with probability 1*

$$\sum_{m^{1/3} < j \le m-m^{1/3}} (Y_{j,m} - \bar{Y}_m)\bar{\varepsilon}_j^{(m)}(t) = o_P\left( m^{1/\alpha}\omega_m^{1/2-1/\alpha} \right). \tag{25}$$

*The statements of the theorem remain valid for selection with replacement, except that in this case the $\delta_j^{(m)}(t)$, $j = 1,\ldots,m$ are independent $B(\lfloor mt \rfloor, 1/m)$ random variables.*

Letting $\bar{\delta}_j^{(m)}(t) = \delta_j^{(m)}(t) - E\delta_j^{(m)}(t)$ and

$$A_m = \left( \sum_{j=1}^m (Y_{j,m} - \bar{Y}_m)^2 \right)^{1/2} = \sqrt{m}\,\widehat{\sigma}_m \approx m^{1/\alpha}\omega_m^{1/2-1/\alpha},$$

relation (25) shows that the asymptotic behavior of $A_m^{-1}\widehat{Z}_m(t)$ is the same as that of

$$A_m^{-1} \sum_{j \in I_m} (Y_{j,m} - \bar{Y}_m) \bar{\varepsilon}_j^{(m)}(t) \tag{26}$$

with

$$I_m = \{k : 1 \leq k \leq \lfloor m^{1/3} \rfloor \text{ or } m - \lceil m^{1/3} \rceil < k \leq m\}.$$

By (25), the expression in (26) can be replaced by $A_m^{-1} \sum_{j \in I_m} (Y_{j,m} - \bar{Y}_m) \bar{\delta}_j^{(m)}(t)$, a normed sum of i.i.d. random variables. Using a tightness argument and the functional central limit theorem under Ljapunov's condition, we get

**Corollary 3.1.** *Assume $H_0$ and define*

$$\widehat{T}_m = \max_{1 \leq k \leq m} \frac{1}{\sqrt{m}\widehat{\sigma}_m} \sum_{j=1}^{k} (Y_{\pi(j)} - \bar{Y}_m).$$

*Then conditionally on* **X**, *for almost all paths*

$$P_{\mathbf{X}}(\widehat{T}_m \leq x) \longrightarrow P\left(\sup_{0 \leq t \leq 1} |B(t)| \leq x\right) \qquad \text{for } n \to \infty.$$

A similar result holds for Darling-Erdős type functionals:

**Corollary 3.2.** *Assume $H_0$ and define*

$$\widehat{T}_{m,1} := \max_{1 \leq k \leq m} \sqrt{\frac{m}{k(m-k)}} \frac{1}{\widehat{\sigma}_m} \left| \sum_{j=1}^{k} (Y_{\pi(j)} - \bar{Y}_m) \right|.$$

*Then conditionally on* **X**, *for almost all paths*

$$P_{\mathbf{X}}\left((2\log\log m)^{1/2}\widehat{T}_{m,1} \leq x + 2\log\log m + \frac{1}{2}\log\log\log m - \frac{1}{2}\log(\pi)\right)$$
$$\longrightarrow \exp(-2\exp(-x)).$$

Corollaries 3.1 and 3.2 show that trimming and permuting the sample provides a satisfactory setup for change point problems under infinite variance. Just like in case of finite variances (cf. Theorem 3.1), Corollary 3.2 remains true under the change–point alternative. A small simulation study is given below, showing simulated critical values and the empirical power of the trimmed tests.

Consider the location model in (22) with

$$n \in \{100, 200\}, K \in \{n/4, n/2, 3n/4\}, \mu = 0, \delta \in \{0, 2, 4\}$$

and with i.i.d. errors $e_j$ having distribution function

$$F(x) = \begin{cases} \frac{1}{2}(1-x)^{-1.5} & \text{for } x \leq 0 \\ 1 - \frac{1}{2}(1+x)^{-1.5} & \text{for } x < 0. \end{cases}$$

We use trimming with $\omega_n = \lfloor n^\beta \rfloor$, $\beta \in \{0.2, 0.3, 0.4\}$. To simulate the critical values we generate a random sample $(X_1, \ldots, X_n)$ according to the above model and trim it to obtain $(Y_1, \ldots, Y_m)$. For $N = 10^5$ permutations of the integers $\{1, \ldots, m\}$ we calculate the values of $\widehat{T}_m$ and $\widehat{T}_{m,1}$ defined in Corollaries 3.1 and 3.2, respectively. The computation of the empirical quantiles yields the desired critical values. Tables 1 and 3 summarize our results for $\widehat{T}_m$ and $\widehat{T}_{m,1}$, respectively.

**Table 1** Simulated quantiles of $\widehat{T}_m$

| | | | $\beta = 0.2$ | | | $\beta = 0.3$ | | | $\beta = 0.4$ | | |
|---|---|---|---|---|---|---|---|---|---|---|---|
| $n$ | $K$ | $\delta$ | 10 % | 5 % | 1 % | 10 % | 5 % | 1 % | 10 % | 5 % | 1 % |
| 100 | - | 0 | 1.115 | 1.225 | 1.429 | 1.144 | 1.271 | 1.514 | 1.142 | 1.264 | 1.495 |
| 100 | 25 | 2 | 1.127 | 1.239 | 1.467 | 1.157 | 1.288 | 1.541 | 1.153 | 1.283 | 1.535 |
| 100 | 25 | 4 | 1.144 | 1.264 | 1.501 | 1.161 | 1.292 | 1.542 | 1.164 | 1.295 | 1.556 |
| 100 | 50 | 2 | 1.137 | 1.258 | 1.489 | 1.150 | 1.280 | 1.523 | 1.159 | 1.289 | 1.539 |
| 100 | 50 | 4 | 1.154 | 1.276 | 1.514 | 1.159 | 1.289 | 1.539 | 1.167 | 1.298 | 1.553 |
| 100 | 75 | 2 | 1.133 | 1.255 | 1.483 | 1.153 | 1.281 | 1.528 | 1.155 | 1.283 | 1.536 |
| 100 | 75 | 4 | 1.148 | 1.276 | 1.514 | 1.163 | 1.293 | 1.548 | 1.159 | 1.291 | 1.550 |
| 200 | - | 0 | 1.166 | 1.293 | 1.549 | 1.144 | 1.268 | 1.498 | 1.169 | 1.303 | 1.560 |
| 200 | 50 | 2 | 1.176 | 1.306 | 1.561 | 1.157 | 1.284 | 1.525 | 1.183 | 1.314 | 1.577 |
| 200 | 50 | 4 | 1.179 | 1.311 | 1.566 | 1.172 | 1.300 | 1.549 | 1.182 | 1.313 | 1.582 |
| 200 | 100 | 2 | 1.168 | 1.300 | 1.551 | 1.171 | 1.298 | 1.553 | 1.169 | 1.299 | 1.558 |
| 200 | 100 | 4 | 1.179 | 1.307 | 1.561 | 1.182 | 1.314 | 1.576 | 1.181 | 1.315 | 1.574 |
| 200 | 150 | 2 | 1.149 | 1.270 | 1.503 | 1.154 | 1.279 | 1.530 | 1.176 | 1.307 | 1.569 |
| 200 | 150 | 4 | 1.164 | 1.289 | 1.531 | 1.158 | 1.284 | 1.537 | 1.181 | 1.309 | 1.569 |

Note that the differences between the estimated quantiles under the null ($\delta = 0$) and under the alternative hypotheses are small, just as in case of finite variances in Antoch and Hušková [1]. Comparing the simulated values with the asymptotic ones given in Tables 2 and 4, one can note relatively large differences for $n = 100$ (in particular in the case of $\widehat{T}_{m,1}$).

**Table 2** Asymptotic critical values of $T_m$

| 10 % | 5 % | 1 % |
|---|---|---|
| 1.224 | 1.358 | 1.628 |

The quantiles of $\widehat{T}_{m,1}$ show more fluctuations than those of $\widehat{T}_m$. Note that increasing $\beta$ will stabilize the simulated quantiles.

**Table 3** Simulated quantiles of $\widehat{T}_{m,1}$

| | | | $\beta = 0.2$ | | | $\beta = 0.3$ | | | $\beta = 0.4$ | | |
|---|---|---|---|---|---|---|---|---|---|---|---|
| $n$ | $K$ | $\delta$ | 10 % | 5 % | 1 % | 10 % | 5 % | 1 % | 10 % | 5 % | 1 % |
| 100 | - | 0 | 2.835 | 3.174 | 3.961 | 2.900 | 3.199 | 3.540 | 2.719 | 2.945 | 3.453 |
| 100 | 25 | 2 | 2.793 | 3.143 | 3.511 | 2.827 | 3.032 | 3.465 | 2.661 | 2.898 | 3.371 |
| 100 | 25 | 4 | 2.719 | 2.945 | 3.415 | 2.694 | 2.946 | 3.441 | 2.660 | 2.898 | 3.385 |
| 100 | 50 | 2 | 2.939 | 3.328 | 3.960 | 2.703 | 2.978 | 3.386 | 2.650 | 2.881 | 3.335 |
| 100 | 50 | 4 | 2.747 | 3.034 | 3.625 | 2.640 | 2.872 | 3.342 | 2.635 | 2.869 | 3.346 |
| 100 | 75 | 2 | 2.926 | 3.339 | 3.808 | 2.875 | 3.353 | 3.765 | 2.764 | 3.012 | 3.438 |
| 100 | 75 | 4 | 2.864 | 3.091 | 3.667 | 2.731 | 2.950 | 3.466 | 2.685 | 2.921 | 3.411 |
| 200 | - | 0 | 3.065 | 3.431 | 4.317 | 3.065 | 3.469 | 4.560 | 2.924 | 3.286 | 3.994 |
| 200 | 50 | 2 | 3.035 | 3.393 | 4.501 | 3.009 | 3.433 | 4.527 | 2.896 | 3.243 | 3.844 |
| 200 | 50 | 4 | 2.959 | 3.242 | 4.071 | 2.929 | 3.249 | 4.078 | 2.851 | 3.110 | 3.541 |
| 200 | 100 | 2 | 3.027 | 3.406 | 4.500 | 2.983 | 3.317 | 4.004 | 2.868 | 3.179 | 3.775 |
| 200 | 100 | 4 | 2.886 | 3.188 | 3.799 | 2.892 | 3.112 | 3.751 | 2.771 | 3.019 | 3.508 |
| 200 | 150 | 2 | 3.024 | 3.519 | 5.092 | 2.951 | 3.356 | 4.125 | 2.832 | 3.070 | 3.570 |
| 200 | 150 | 4 | 2.949 | 3.323 | 4.386 | 2.874 | 3.183 | 3.812 | 2.802 | 3.051 | 3.547 |

**Table 4** Asymptotic critical values of $T_{m,1}$

| $n$ | $\beta$ | 10 % | 5 % | 1 % |
|---|---|---|---|---|
| 100 | .2 | 3.223 | 3.636 | 4.572 |
| 100 | .3 | 3.222 | 3.636 | 4.572 |
| 100 | .4 | 3.218 | 3.634 | 4.575 |
| 200 | .2 | 3.264 | 3.658 | 4.552 |
| 200 | .3 | 3.262 | 3.658 | 4.552 |
| 200 | .4 | 3.260 | 3.656 | 4.553 |

Figures 2–5 show the empirical power of the test based on the (non permuted) test statistics $T_m$, $T_{m,1}$ for each $\delta \in \{-3, -2.9, \ldots, 2.9, 3\}$.

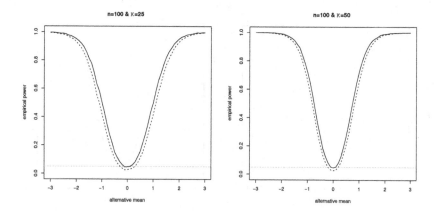

**Fig. 2** Empirical power of $T_m$ with empirical (dashed) and asymptotic (solid) critical values, $\alpha = 0.05, n = 100, \beta = 0.3$ and $K = 25, 50$ (left, right)

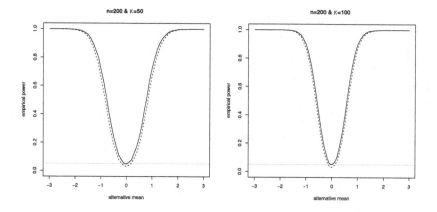

**Fig. 3** Empirical power of $T_m$ with empirical (dashed) and asymptotic solid) critical values, $\alpha = 0.05, n = 200, \beta = 0.3$ and $K = 50, 100$ (left, right)

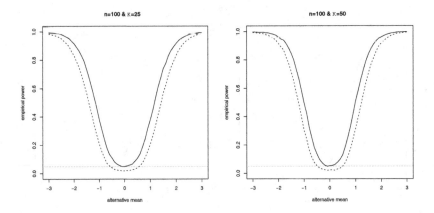

**Fig. 4** Empirical power of $T_{m,1}$ with empirical (dashed) and asymptotic (solid) critical values $\alpha = 0.05, n = 100, \beta = 0.3$ and $K = 25, 50$ (left, right)

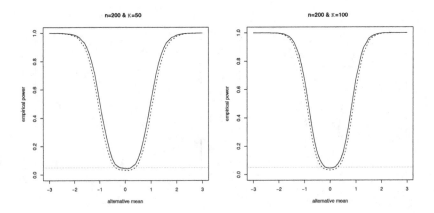

**Fig. 5** Empirical power of $T_{m,1}$ with empirical (dashed) and asymptotic (solid) critical values $\alpha = 0.05, n = 200, \beta = 0.3$ and $K = 50, 100$ (left, right)

The figures show that the test based on $T_{m,1}$ is conservative when we use the asymptotic critical values.

In conclusion we note that the type of trimming we used above is not the only possibility to eliminate the large elements of the sample $(X_1, \ldots, X_n)$. Alternatively, we can remove from the sample the $\omega_n$ elements with the largest absolute values. In [7] we determined the asymptotic distribution of permuted and bootstrapped CUSUM statistics under this kind of trimming. While the limit distribution of $\widehat{T}_m$ remains the same in this case, note that, surprisingly, the asymptotic theory of the two trimming procedures is different, see [7], [12], [13] for further information.

**Acknowledgements** István Berkes' research was supported by FWF grant S9603-N23 and OTKA grants K 61052 and K 67961, Lajos Horváth's research was partially supported by NSF grant DMS 00905400 and Johannes Schauer's research was partially supported by FWF grant S9603-N23.

# References

[1] Antoch, J., and Hušková, M. (2001). Permutation tests in change point analysis. *Stat. Prob. Letters* **53**, 37–46.

[2] Athreya, K. (1987). Bootstrap of the mean in the infinite variance case. *Ann. Statist.* **15**, 724–731.

[3] Aue, A., Berkes, I., and Horváth, L. (2008). Selection from a stable box. *Bernoulli* **14**, 125–139.

[4] Bergström, H. (1971). The limit problem for sums of independent random variables which are not uniformly asymptotically negligible. Transactions of the Sixth Prague Conference on Information Theory, Statistical Decision Functions, Random Processes. Tech. Univ. Prague, 125–135. Academia, Prague.

[5] Berkes, I., Horváth, L., Hušková, M., and Steinebach, J. (2004). Applications of permutations to the simulations of critical values. *J. of Nonparametric Statistics* **16**, 197–216.

[6] Berkes, I., Horváth, L., and Schauer, J. (2009). Non-central limit theorems for random selections. *Prob. Theory Rel. Fields*, to appear.

[7] Berkes, I., Horváth, L., and Schauer, J. (2009). Asymptotics of trimmed CUSUM statistics. *Bernoulli*, submitted.

[8] Bickel, P., and Freedman, D. (1981). Some asymptotic theory for the bootstrap. *Ann. Statist.* **9**, 1196–1217.

[9] Csörgő, M., and Horváth, L. (1997). Limit theorems in change-point analysis. Wiley, New York.

[10] Csörgő, S. (2002). Rates of merge in generalized St. Petersburg games. *Acta Sci. Math. (Szeged)* **68**, 815–847.

[11] Darling, D. (1952). The influence of the maximum term in the addition of independent random variables. *Trans. Amer. Math. Soc.* **73**, 95–107.

[12] Griffin, P. S., and Pruitt, W. E. (1987). The central limit problem for trimmed
     sums. *Math. Proc. Cambridge Philos. Soc.* **102**, 329–349.

[13] Griffin, P. S., and Pruitt, W. E. (1989). Asymptotic normality and subsequential
     limits of trimmed sums. *Ann. Probab.* **17**, 1186–1219.

[14] Hušková, M. (2004). Permutation principle and bootstrap change point analy-
     sis. In: *Asymptotic Methods in Stochastics*, eds. L. Horváth and B. Szyszkow-
     icz. American Mathematical Society, Providence, 273–291.

[15] Rosén, B. (1965). Limit theorems for sampling from finite populations. *Ark.
     Mat.* **5**, 383–424.

[16] Skorokhod, A. V. (1956). Limit theorems for stochastic processes. (Russian)
     *Teor. Veroyatnost. i Primenen.* **1**, 289–319.

# Max–Stable Processes: Representations, Ergodic Properties and Statistical Applications

Stilian A. Stoev

**Abstract** Max–stable processes arise as limits in distribution of component–wise maxima of independent processes, under suitable centering and normalization. Therefore, the class of max–stable processes plays a central role in the study and modeling of extreme value phenomena. This chapter starts with a review of classical and some recent results on the representations of max–stable processes. Recent results on necessary and sufficient conditions for the ergodicity and mixing of stationary max–stable processes are then presented. These results readily yield the consistency of many statistics for max–stable processes. As an example, a new estimator of the extremal index for a stationary max–stable process is introduced and shown to be consistent.

## 1 Introduction

In the past 30 years, the structure of max–stable random vectors and processes has been vigorously explored. A number of seminal papers such as Balkema and Resnick [1], de Haan [4, 5], de Haan and Pickands [9], Gine, Hahn and Vatan [13], Resnick and Roy [20], just to name a few, have lead to an essentially complete picture of the dependence structure of max–stable processes. Complete accounts of the state-of-the-art can be found in the books of Resnick [18], de Haan and Ferreira [6], and Resnick [19], and the references therein.

The stochastic process $X$ is said to be max–stable if all its finite–dimensional distributions are max–stable. Recall that a random vector $\mathbf{Y} = (Y(j))_{1 \le j \le d}$ in $\mathbb{R}^d$ is said to be *max–stable* if, for all $n \in \mathbb{N}$, there exist $\mathbf{a}_n > 0$ and $\mathbf{a}_n, \mathbf{b}_n \in \mathbb{R}^d$, such that

Stilian A. Stoev
Department of Statistics, The University of Michigan,
439 W. Hall, 1085 S. University, Ann Arbor, MI 48109–1107, e-mail: sstoev@umich.edu

P. Doukhan et al. (eds.), *Dependence in Probability and Statistics*,
Lecture Notes in Statistics 200, DOI 10.1007/978-3-642-14104-1_2,
© Springer-Verlag Berlin Heidelberg 2010

$$\bigvee_{1 \le i \le n} \mathbf{Y}_i \overset{d}{=} \mathbf{a}_n \mathbf{Y} + \mathbf{b}_n.$$

Here $\mathbf{Y}_i = (Y_i(j))_{1 \le j \le d}$, $i = 1, \dots, n$ are independent copies of $\mathbf{Y}$ and the above inequalities, vector multiplications, additions and maxima are taken coordinate–wise.

The importance of max–stable processes in applications stems from the fact that they appear as limits of component–wise maxima. Suppose that $\xi_i = \{\xi_i(t)\}_{t \in T}$ are independent and identically distributed stochastic processes, where $T$ is an arbitrary index set. Consider the coordinate–wise maximum process

$$M_n(t) := \bigvee_{1 \le i \le n} \xi_i(t) \equiv \max_{1 \le i \le n} \xi_i(t), \ t \in T.$$

Suppose that for suitable non–random sequences $a_n(t) > 0$ and $b_n(t)$, $t \in T$, we have

$$\left\{ \frac{1}{a_n(t)} M_n(t) - b_n(t) \right\}_{t \in T} \overset{f.d.d.}{\longrightarrow} \{X(t)\}_{t \in T}, \tag{1}$$

as $n \to \infty$, for some *non–degenerate* limit process $X = \{X(t)\}_{t \in T}$, where $\overset{f.d.d.}{\longrightarrow}$ denotes convergence of the finite–dimensional distributions. The processes that appear in the limit of (1) are *max–stable* (see e.g. Proposition 5.9 in [18]). The classical results of Fisher & Tippett and Gnedenko indicate that the marginal distributions of $X$ are one of three types of *extreme value distributions*: Fréchet, Gumbel, or reversed Weibul. The dependence structure of the limit $X$, however, can be quite intricate. Our main focus here is on the study of various aspects of the dependence structure of max–stable processes.

Max–stable processes have a peculiar property, namely their dependence structure is in a sense invariant to the type of their marginals. More precisely, consider a process $X = \{X(t)\}_{t \in T}$ and its transformed version $h \circ X = \{h_t(X_t)\}_{t \in T}$, where $h = \{h_t(\cdot)\}_{t \in T}$ is a collection of deterministic functions, strictly increasing on their domains. It turns out that if $X$ is a max–stable process and if the marginals of $h \circ X$ are extreme value distributions, then the transformed process $h \circ X$ is also max–stable (see e.g. Proposition 5.10 in [18]). That is, one does not encounter max–stable processes with more rich dependence structures if one allows for the marginal distributions of $X$ to be of different types. Thus, for convenience and without loss of generality, we shall focus here on max–stable processes $X = \{X(t)\}_{t \in T}$ with $\alpha$–Fréchet marginals. A random variable $\xi$ is said to have the $\alpha$–Fréchet distribution if:

$$\mathscr{P}\{\xi \le x\} = \exp\{-\sigma^\alpha x^{-\alpha}\}, \ (x > 0),$$

for some $\sigma > 0$ and $\alpha > 0$. The parameter $\sigma > 0$ plays the role of a scale coefficient, and thus, by analogy with the convention for sum–stable processes, we shall use the notation

$$\|\xi\|_\alpha := \sigma.$$

Note that here $\|\cdot\|_\alpha$ is not the usual $L^\alpha$–norm but we have $\|c\xi\|_\alpha = c\|\xi\|_\alpha$, for all $c > 0$. The $\alpha$–Fréchet laws have heavy Pareto–like tails with tail exponent $\alpha > 0$, that is,

$$\mathscr{P}\{\xi > x\} \sim \|\xi\|_\alpha^\alpha x^{-\alpha}, \quad \text{as } x \to \infty.$$

Therefore, the $p$–moment $(p > 0)$, $\mathbb{E}\xi^p < \infty$ is finite if and only if $p < \alpha$.

It is convenient to introduce the notion of an $\alpha$–Fréchet process. Namely, the process $X = \{X(t)\}_{t \in T}$ is said to be an $\alpha$–Fréchet process if all (positive) max–linear combinations of $X(t)$'s:

$$\bigvee_{1 \le i \le k} a_i X(t_i), \quad a_i \ge 0, t_i \in T,$$

are $\alpha$–Fréchet random variables.

It turns out that the max–stable processes with $\alpha$–Fréchet marginals are precisely the $\alpha$–Fréchet processes (see, de Haan [4]). Therefore, in the sequel we shall use the terms *Fréchet processes* and *max–stable processes with Fréchet marginals* interchangeably.

Let now $X$ be an $\alpha$–Fréchet process. The structure of the finite–dimensional distributions of $X$ is already known, In fact, we have the following explicit formula of the finite–dimensional distributions of $X$:

$$\mathscr{P}\{X_{t_i} \le x_i, \ 1 \le i \le k\} = \exp\left\{ -\int_0^1 \bigvee_{1 \le i \le k} \left(\frac{f_{t_i}(u)}{x_i}\right)^\alpha du \right\}, \quad (x_i > 0, \ 1 \le i \le k), \quad (2)$$

where $f_{t_i}(u) \ge 0$ are suitable Borel functions, such that $\int_0^1 f_{t_i}^\alpha(u)du < \infty$, for $1 \le i \le k$. The $f_{t_i}(u)$'s are known as *spectral functions* of the max–stable vector $(X_{t_i})_{1 \le i \le k}$ and even though they are not unique, they will play an important role in our representations of max–stable processes. Observe for example, that (2) yields

$$\mathscr{P}\left\{ \bigvee_{1 \le i \le k} a_i X_{t_i} \le x \right\} = \exp\left\{ -\int_0^1 \left( \bigvee_{1 \le i \le k} a_i f_{t_i}(u) \right)^\alpha du\, x^{-\alpha} \right\},$$

and therefore $\bigvee_{1 \le i \le k} a_i X_{t_i}$ is an $\alpha$–Fréchet variable with scale coefficient

$$\left\| \bigvee_{1 \le i \le k} a_i X_{t_i} \right\|_\alpha^\alpha = \int_0^1 \left( \bigvee_{1 \le i \le k} a_i f_{t_i}(u) \right)^\alpha du.$$

Thus, the knowledge of the spectral functions $\{f_t(u)\}_{t \in T} \subset L_+^\alpha([0,1], du)$ allows us to handle all finite–dimensional distributions of the process $X$.

One can alternatively express the finite–dimensional distributions in (2) by using the *spectral measure* of the vector $(X_{t_i})_{1 \le i \le k}$. Namely, consider an arbitrary norm $\|\cdot\|$ in $\mathbb{R}^k$ and let $\mathbb{S}_+ := \{w = (w_i)_{1 \le i \le k} : w_i \ge 0, \|w\| = 1\}$ be the *non–negative unit sphere* in $\mathbb{R}^k$. We then have

$$\mathscr{P}\{X_{t_i} \le x_i, \ 1 \le i \le k\} = \exp\left\{ -\int_{\mathbb{S}_+} \bigvee_{1 \le i \le k} \frac{w_i}{x_i^\alpha} \nu_{\mathbb{S}_+}(dw) \right\}, \quad (3)$$

where $v_{\mathbb{S}_+}(dw)$ is a finite measure on $\mathbb{S}_+$.

The two types of representations in (2) and (3) have both advantages and disadvantages depending on the particular setting. The finite measure $v_{\mathbb{S}_+}$ associated with the max–stable vector $(X_{t_i})_{1 \le i \le k}$ is said to be its *spectral measure* and it is uniquely determined. Thus, when handling max–stable random vectors of fixed dimension, the spectral measure is a natural object to use and estimate statistically. On the other hand, when handling stochastic processes, one encounters spectral measures defined on spaces of different dimensions, which may be hard to reconcile. In such a setting, it may be more natural to use representations based on a set of spectral functions $\{f_t\}_{t \in T}$, which are ultimately defined on the same measure space.

More details and the derivations of (2) and (3) can be found in [18]. Novel perspectives to spectral measures on 'infinite dimensional' spaces are adopted in Gine, Hahn and Vatan [13], de Haan and Lin [7, 8] and de Haan and Ferreira [6]. Hult and Lindskog [14] develop powerful new tools based on the related notion of regular variation in infinite–dimensional function spaces.

Let now $X = \{X(t)\}_{t \in T}$ with $T = \mathbb{R}$ or $\mathbb{Z}$ be a stationary $\alpha$–Fréchet process. From statistical perspective, it is important to know whether the process $X$ is ergodic, mixing, or non–ergodic. Despite the abundance of literature on max–stable processes, the problem of ergodicity had not been explored until recently. To the best of our knowledge only Weintraub in [27] addressed it indirectly by introducing mixing conditions through certain measures of dependence. Recently, in [25], by following the seminal work of [3], we obtained necessary and sufficient conditions for the process $X$ to be ergodic or mixing. In the case of mixing, these conditions take a simple form and are easy to check for many particular cases of max–stable processes.

*The goal of this chapter is to primarily review results estabilished in [24, 25].* This is done in Sections 2 and 3 below. These results are then illustrated and applied to some statistical problems in Seciton 4. Section 4.2 contains new results on the consistency of extremal index estimators for stationary max–stable processes.

## 2 Representations of Max–Stable Processes

Let $X = \{X(t)\}_{t \in \mathbb{R}}$ be an $\alpha$–Fréchet process ($\alpha > 0$) indexed by $\mathbb{R}$. As indicated above all finite–dimensional distributions of $X$ can be expressed in terms of a family of *spectral functions* in

$$L_+^\alpha([0,1], du) = \{f : [0,1] \to \mathbb{R}_+ : \int_{[0,1]} f^\alpha(u) du < \infty\}.$$

The seminal paper of de Haan [5] shows that provided $X = \{X(t)\}_{\mathbb{R}}$ is continuous in probability, there exists a family of *spectral functions* $\{f_t(u)\}_{t \in \mathbb{R}} \subset L_+^\alpha(du)$ indexed by $\mathbb{R}$, which yield (2). This was done from the appealing perspective of Poisson point processes. Namely, let $X = \{X(t)\}_{t \in \mathbb{R}}$ be continuous in probability. Then, there exist a collection of non–negative functions $\{f_t(u)\}_{t\in\mathbb{R}} \subset L_+^\alpha([0,1], du)$, such that

$$\{X(t)\}_{t\in T} \stackrel{d}{=} \Big\{ \bigvee_{i\in\mathbb{N}} \frac{f_t(U_i)}{\varepsilon_i^{1/\alpha}} \Big\}_{t\in T}, \tag{4}$$

where $\{(U_i, \varepsilon_i)\}_{i\in\mathbb{N}}$ is a Poisson point process on $[0,1] \times [0,\infty]$ with intensity $du \times ds$, and where $\stackrel{d}{=}$ means equality in the sense of finite–dimensional distributions.

We now present an alternative but ultimately equivalent approach to representing $\alpha$–Fréchet max–stable processes developed in Stoev and Taqqu [24]. It is based on the notion of extremal integrals with respect to $\alpha$–Fréchet sup–measures.

**Definition 0.1.** Let $\alpha > 0$ and let $(E, \mathscr{E}, \mu)$ be a measure space. A random set–function $M_\alpha$, defined on $\mathscr{E}$, is said to be an $\alpha$–Fréchet random sup–measure with control measure $\mu$ if the following conditions hold:

*(i)* For all disjoint $A_j \in \mathscr{E}$, $1 \le j \le n$, the random variables $M_\alpha(A_j)$, $1 \le j \le n$ are independent.

*(ii)* For all $A \in \mathscr{E}$, the random variable $M_\alpha(A)$ is $\alpha$–Fréchet, with scale coefficient $\|M_\alpha(A)\|_\alpha = \mu(A)^{1/\alpha}$, i.e.

$$\mathscr{P}\{M_\alpha(A) \le x\} = \exp\{-\mu(A)x^{-\alpha}\}, \quad x > 0. \tag{5}$$

*(iii)* For all disjoint $A_j \in \mathscr{E}$, $j \in \mathbb{N}$,

$$M_\alpha(\cup_{j\in\mathbb{N}}A_j) = \bigvee_{j\in\mathbb{N}} M_\alpha(A_j), \quad \text{almost surely.} \tag{6}$$

By convention, we set $M_\alpha(A) = \infty$ if $\mu(A) = \infty$.

Condition *(i)* in the above definition means that the random measure is *independently scattered* i.e. it assigns independent random variables to disjoint sets and Condition *(ii)* shows that the scale of $M_\alpha(A)$ is governed by the deterministic *control measure* $\mu$ of $M_\alpha$. Relation (6), on the other hand, indicates that the random measure $M_\alpha$ is sup–additive, rather than additive. This is the fundamental difference between the usual additive random measures and the sup–measures. For more general studies of sup–measures see [26]. The important work of Norberg [17] unveils the connections between random sup–measures, the theory of random sets, and random capacities. Here, the focus is on the concrete and simple case of $\alpha$–Fréchet sup–measures, most relevant to the study of max–stable processes.

As shown in Proposition 2.1 of [24] (by using the Kolmogorov's extension theorem) for any measure space $(E, \mathscr{E}, \mu)$ one can construct an $\alpha$–Fréchet random sup–measure $M_\alpha$ with control measure $\mu$, on a sufficiently rich probability space. Given such a random measure $M_\alpha$ on $(E, \mathscr{E}, \mu)$, one can then define the *extremal integral* of a non–negative deterministic function with respect to $M_\alpha$ as follows. Consider first a non–negative simple function $f(u) = \sum_{i=1}^n a_i 1_{A_i}(u)$, $a_i \ge 0$ with disjoint $A_i$'s and define the extremal integral of $f$ as

$$I(f) \equiv \int_E^{ef} f dM_\alpha := \bigvee_{i=1}^n a_i M_\alpha(A_i),$$

i.e. the sum in the typical definition of an integral is replaced by a maximum. Since the $M_\alpha(A_i)$'s are independent and $\alpha$−Fréchet, Relation (5) implies

$$\mathscr{P}\{I(f) \leq x\} = \exp\left\{-\int_E f^\alpha d\mu \, x^{-\alpha}\right\}, \quad x > 0.$$

The following properties are immediate (see e.g. Proposition 2.2 in [24]):

**Properties:**

- For all non−negative simple functions $f$, the extremal integral $\int_E^{ef} f dM_\alpha$ is an $\alpha$−Fréchet random variable with scale coefficient

$$\left\| \int_E^{ef} f dM_\alpha \right\|_\alpha = \left( \int_E f^\alpha d\mu \right)^{1/\alpha}. \tag{7}$$

- *(max−linearity)* For all $a, b \geq 0$ and all non−negative simple functions $f$ and $g$, we have

$$\int_E^{ef} (af \vee bg) dM_\alpha = a \int_E^{ef} f dM_\alpha \vee b \int_E^{ef} g dM_\alpha, \quad \text{almost surely.} \tag{8}$$

- *(independence)* For all simple functions $f$ and $g$, $\int_E^{ef} f dM_\alpha$ and $\int_E^{ef} g dM_\alpha$ are independent *if and only if* $fg = 0$, $\mu$−almost everywhere.

Relation (8) shows that the extremal integrals are *max−linear*. Note that for any collection of non−negative simple functions $f_i$ and $a_i \geq 0$, $1 \leq i \leq n$, we have that

$$\bigvee_{1 \leq i \leq n} a_i \int_E^{ef} f_i dM_\alpha = \int_E^{ef} \left( \bigvee_{1 \leq i \leq n} a_i f_i \right) dM_\alpha$$

is $\alpha$−Fréchet. This shows that the set of extremal integrals of non−negative simple functions is *jointly $\alpha$−Fréchet*, i.e. the distribution of $(I(f_i))_{1 \leq i \leq n}$ is multivariate max−stable. It turns out that one can metrize the convergence in probability in the spaces of jointly $\alpha$−Fréchet random variables by using the following metric:

$$\rho_\alpha(\xi, \eta) := 2\|\xi \vee \eta\|_\alpha^\alpha - \|\xi\|_\alpha^\alpha - \|\eta\|_\alpha^\alpha. \tag{9}$$

If now $\xi = \int_E^{ef} f dM_\alpha$ and $\eta = \int_E^{ef} g dM_\alpha$, for some simple functions $f \geq 0$ and $g \geq 0$, we obtain

$$\rho_\alpha(\xi, \eta) = 2\int_E (f^\alpha \vee g^\alpha) d\mu - \int_E (f^\alpha \vee g^\alpha) d\mu \int_E (f^\alpha \vee g^\alpha) d\mu \equiv \int_E |f^\alpha - g^\alpha| d\mu. \tag{10}$$

By using this relationship one can extend the definition of the extremal integral $\int_E^{ef} f dM_\alpha$ to integrands in the space $L_+^\alpha(\mu) \equiv L_+^\alpha(E, \mathscr{E}, \mu)$ of all non−negative deterministic $f$'s with $\int_E f^\alpha d\mu < \infty$. Moreover, the above properties of the extremal integrals remain valid for all such integrands.

To complete the picture, consider the space

$$\mathcal{M}_\alpha = \overline{\vee - \text{span}}^P \{M_\alpha(A) : A \in \mathcal{E}\}$$

of jointly $\alpha$–Fréchet variables containing all max–linear combinations of $M_\alpha(A)$'s and their limits in probability. One can show that $(\mathcal{M}_\alpha, \rho_\alpha)$ is a complete metric space and $\rho_\alpha$ in (9), as indicated above, metrizes the convergence in probability. Let also $L_+^\alpha(\mu)$ be equipped with the metric

$$\rho_\alpha(f,g) := \int_E |f^\alpha - g^\alpha| d\mu. \tag{11}$$

Then, relation (10) implies that the extremal integral

$$I : L_+^\alpha(\mu) \to \mathcal{M}_\alpha$$

is a *max–linear isometry* between the metric spaces $(L_+^\alpha(\mu), \rho_\alpha)$ and $(\mathcal{M}_\alpha, \rho_\alpha)$, which is one-to-one and onto. Thus, in particular if $\xi_n := \int_E^e f_n dM_\alpha$ and $\xi = \int_E^e f dM_\alpha$, $f_n, f \in L_+^\alpha(\mu)$, we have that

$$\xi_n \overset{P}{\longrightarrow} \xi, \quad \text{as } n \to \infty, \quad \textit{if and only if} \quad \rho_\alpha(f_n, f) = \int_E |f_n^\alpha - f^\alpha| d\mu \longrightarrow 0, \quad \text{as } n \to \infty.$$

For more details see Stoev and Taqqu [24].

The so developed extremal integrals provide us with tools to construct and handle max–stable processes. Indeed, for any collection of deterministic integrands $\{f_t\}_{t \in T} \subset L_+^\alpha(\mu)$, one can define

$$X(t) := \int_E^e f_t dM_\alpha, \quad t \in T. \tag{12}$$

The resulting process $X = \{X(t)\}_{t \in T}$ is $\alpha$–Fréchet and in view of (7) and (8), we obtain

$$\left\| \bigvee_{1 \le i \le k} a_i X(t_i) \right\|_\alpha = \left( \int_E \bigvee_{1 \le i \le k} a_i^\alpha f_{t_i}^\alpha d\mu \right)^{1/\alpha},$$

where $a_i \ge 0$. Therefore, with $a_i := 1/x_i$, $1 \le i \le k$, we obtain

$$\mathscr{P}\{X(t_i) \le x_i, \ 1 \le i \le k\} = \exp\left\{ - \int_E \bigvee_{1 \le i \le k} \frac{f_{t_i}^\alpha}{x_i^\alpha} d\mu \right\}.$$

This shows that the $f_t$'s play the role of the spectral functions of the max–stable process $X$ as in (2) but now these functions can be defined over an arbitrary measure space $(E, \mathcal{E}, \mu)$. Thus, by choosing suitable families of integrands (kernels) $f_t$'s, one can explicitly model and manipulate a variety of max–stable processes. For example, if $E \equiv \mathbb{R}$ is the real line equipped with the Lebesgue measure, one can define the *moving maxima* processes:

$$X(t) := \int_{\mathbb{R}}^{e} f(t-u)M_\alpha(du), \ t \in \mathbb{R}, \tag{13}$$

where $f \geq 0$, $\int_{\mathbb{R}} f^\alpha(u)du < \infty$, and where $M_\alpha$ is an $\alpha$–Fréchet random sup–measure with the Lebesgue control measure. More generally, we define a *mixed moving maxima* process or field as follows:

$$X(t) \equiv X(t_1, \cdots, t_d) := \int_{\mathbb{R}^d \times V}^{e} f(t-u,v)M_\alpha(du,dv), \ t = (t_i)_{i=1}^d \in \mathbb{R}^d, \tag{14}$$

where $f \geq 0$, $\int_{\mathbb{R}^d \times V} f^\alpha(u,v)du\nu(dv) < \infty$ and where now the random sup–measure $M_\alpha$ is defined on the product space $\mathbb{R}^d \times V$ and has control measure $du \times \nu(dv)$, for some measure $\nu(dv)$ on the set $V$.

Further, interesting classes of processes are obtained when the measure space $(E, \mathscr{E}, \mu)$ is viewed as another *probability space* and the collection of deterministic integrands $\{f_t\}_{t \in T}$ is then interpreted as a stochastic process on this probability space. This leads to certain *doubly stochastic* max–stable processes, whose dependence structure is closely related to the stochastic properties of the integrands $f_t$'s. For more details, see Section 4.1 below.

Let $X = \{X(t)\}_{t \in T}$ be an $\alpha$–Fréchet process. As shown in [24], the representation in (4) (or equivalently in (12) with $(E, \mathscr{E}, \mu) \equiv ((0,1), \mathscr{B}_{(0,1)}, dx))$ is possible if and only if the process $X$ is *separable in probability*. The max–stable process $X$ is said to be separable in probability if, there exists a countable set $\mathscr{I} \subset T$, such that for all $t \in T$, the random variable $X_t$ is a limit in probability of max–linear combinations of the type $\max_{1 \leq i \leq n} a_i X_{s_i}$, with $s_i \in \mathscr{I}$ and $a_i \geq 0$, $1 \leq i \leq n$. Clearly, if $T \equiv \mathbb{R}$ and $X$ is continuous in probability, then it is also separable in probability and therefore it has the representation (4) with suitable $f_t$'s (see Theorem 3 in [5]). On the other hand, even if $X$ is not separable in probability, it may still be possible to express as in (12) provided that the measure space $(E, \mathscr{E}, \mu)$ is sufficiently rich.

**Remarks:**

1. The representation (4) is similar in spirit to the Le Page, Woodroofe & Zinn's series representation for sum–stable processes. Namely, let $X = \{X(t)\}_{t \in \mathbb{R}}$ be an $\alpha$–stable process, which is separable in probability. For simplicity, suppose that $X$ is totally skewed to the right and such that $0 < \alpha < 1$. Then, by Theorems 3.10.1 and 13.2.1 in Samorodnitsky and Taqqu [22], we have

$$\{X(t)\}_{t \in \mathbb{R}} \stackrel{d}{=} \left\{ \sum_{i \in \mathbb{N}} \frac{f_t(U_i)}{\varepsilon_i^{1/\alpha}} \right\}_{t \in \mathbb{R}}, \tag{15}$$

where $\{f_t(u)\}_{t \in \mathbb{R}} \subset L^\alpha([0,1], du)$, and $\{(U_i, \varepsilon_i)\}_{i \in \mathbb{N}}$ is a standard Poisson point process on $[0,1] \times [0, \infty]$. Relation (15) is analogous to (4) where the sum is replaced by a maximum and only non–negative spectral functions $f_t(\cdot)$'s are considered.

2. The representation (4) is particularly convenient when studying the path properties of max–stable processes. It was used in [20] to establish necessary and sufficient conditions for the continuity of the paths of max–stable processes.
3. The moving maxima (M2) (in discrete time) were first considered by Deheuvels [11]. Zhang and Smith [29] studied further the discrete–time multivariate mixed moving maxima (M4 processes) generated by sequences of independent $\alpha$–Fréchet variables.

## 3 Ergodic Properties of Stationary Max–stable Processes

Let $X = \{X(t)\}_{t \in \mathbb{R}}$ be a (strictly) stationary $\alpha$–Fréchet process as in (12). To be able to discuss ergodicity in continuous time, we shall suppose that $X$ is measurable. This is not a tall requirement since any continuous in probability process has a measurable modification. All results are valid and in fact have simpler versions in discrete time. We first recall the definitions of ergodicity and mixing in our context.

One can introduce a group of *shift operators* $S_\tau$, $\tau \in \mathbb{R}$, which acts on all random variables, measurable with respect to $\{X(t)\}_{t \in \mathbb{R}}$. Namely, for all $\xi = g(X_{t_1}, \cdots, X_{t_k})$, we define

$$S_\tau(\xi) := g(X_{\tau+t_1}, \cdots, X_{\tau+t_k}),$$

where $g : \mathbb{R}^k \to \mathbb{R}$ is a Borel function. The definition of the $S_\tau$'s can be extended to the class of all $\{X_t\}_{t \in \mathbb{R}}$–measurable random variables. Note also that $S_t \circ S_s = S_{t+s}$, $t, s \in \mathbb{R}$. Clearly, the shift operators map indicator functions to indicator functions and therefore one can define $S_\tau(A) := \{S_\tau(1_A) = 1\}$, for all events $A \in \sigma\{X_t, t \in \mathbb{R}\}$. These mappings are well–defined and unique up to equality almost surely (for more details, see e.g. Ch. IV in [21]).

The stationarity of the process $X$ implies that the shifts $S_\tau$'s are *measure preserving*, i.e.

$$\mathscr{P}(S_\tau(A)) = \mathscr{P}(A), \quad \text{for all } A \in \sigma\{X_t, t \in \mathbb{R}\}.$$

Let now $\mathscr{F}_{inv}$ denote the $\sigma$–algebra of *shift–invariant* sets, namely, the collection of all $A \in \sigma\{X_t, t \in \mathbb{R}\}$ such that $\mathscr{P}(A \Delta S_\tau(A)) = 0$ for all $\tau \in \mathbb{R}$.

Recall that the process $X$ is said to be *ergodic* if the shift–invariant $\sigma$–algebra $\mathscr{F}_{inv}$ is trivial, i.e. for all $A \in \mathscr{F}_{inv}$, we have that either $\mathscr{P}(A) = 0$ or $\mathscr{P}(A) = 1$. On the other hand, $X$ is said to be *mixing* if

$$\mathscr{P}(A \cap S_\tau(B)) \longrightarrow \mathscr{P}(A)\mathscr{P}(B), \quad \text{as } \tau \to \infty,$$

for all $A, B \in \sigma\{X_t, t \in \mathbb{R}\}$.

It is easy to show that mixing implies ergodicity. Furthermore, ergodicity has important statistical implications. Indeed, fix $t_i \in \mathbb{R}$, $1 \leq i \leq k$ and let $h : \mathbb{R}^k \to \mathbb{R}$ be a Borel measurable function such that $\mathbb{E}|h(X(t_1), \cdots, X(t_k))| < \infty$. The Birgkhoff's ergodic theorem implies that, as $T \to \infty$,

$$\frac{1}{T}\int_0^T h(X(\tau+t_1),\cdots,X(\tau+t_k))d\tau \longrightarrow \xi,$$

almost surely and in the $L^1$−sense, where $\mathbb{E}\xi = \mathbb{E}h(X(t_1),\cdots,X(t_k))$. The limit $\xi$ is shift–invariant, that is $S_\tau(\xi) = \xi$, almost surely, for all $\tau > 0$, and therefore $\xi$ is measurable with respect to $\mathscr{F}_{inv}$. Hence, if the process $X$ is ergodic, then the limit $\xi$ is constant, and we have the following strong law of large numbers:

$$\frac{1}{T}\int_0^T h(X(\tau+t_1),\cdots,X(\tau+t_k))d\tau \overset{a.s. \ \& \ L^1}{\longrightarrow} \mathbb{E}h(X(t_1),\cdots,X(t_k)), \quad \text{as } T \to \infty.$$
(16)

In fact, one can show that $X$ is ergodic if and only if Relation (16) holds, for all such Borel functions $h$ and all $k \in \mathbb{N}$. For more details on ergodicity and mixing, see e.g. [21].

Relation (16) indicates the importance of knowing whether a process $X$ is ergodic or not. Ergodicity implies the strong consistency of a wide range of statistics based on the empirical time–averages in (16).

Our goal in this section is to review necessary and sufficient conditions for the ergodicity or mixing of the process $X$. These conditions will be formulated in terms of the deterministic integrands $\{f_t\}_{t\in\mathbb{R}} \subset L_+^\alpha(\mu)$ and the important notion of *max–linear isometry*.

**Definition 0.2.** A mapping $U : L_+^\alpha(\mu) \to L_+^\alpha(\mu)$ is said to be a max–linear isometry, if

*(i)* For all $f,g \in L_+^\alpha(\mu)$ and $a,b \geq 0$,

$$U(af \vee bg) = aU(f) \vee bU(g), \quad \mu\text{−a.e.}$$

*(ii)* For all $f \in L_+^\alpha(\mu)$,

$$\|U(f)\|_{L_+^\alpha(v)} = \|f\|_{L_+^\alpha(\mu)}.$$

Consider a collection of max–linear isometries $U_t : L_+^\alpha(\mu) \to L_+^\alpha(\mu)$, which forms a group with respect to composition, indexed by $t \in \mathbb{R}$, i.e. $U_t \circ U_s = U_{t+s}$, $t,s \in \mathbb{R}$ and $U_0 \equiv \mathrm{id}_E$.

Now, fix $f_0 \in L_+^\alpha(\mu)$, let $f_t := U_t(f_0)$, $t \in \mathbb{R}$, and consider the $\alpha$−Fréchet process

$$X(t) := \int_E^{ef} U_t(f_0)dM_\alpha, \quad t \in \mathbb{R}. \tag{17}$$

Definition 0.2 and the group structure of the $U_t$'s readily implies that $X = \{X(t)\}_{t\in\mathbb{R}}$ is stationary. Indeed,

$$\mathscr{P}\{X(\tau+t_i) \leq x_i, \ 1 \leq i \leq k\} = \exp\left\{-\int_E \bigvee_{1\leq i\leq k} \frac{U_\tau(f_{t_i})^\alpha}{x_i^\alpha}d\mu\right\}$$

$$= \exp\left\{-\int_E U_\tau\left(\bigvee_{1\leq i\leq k} \frac{f_{t_i}a}{x_i^\alpha}\right)^\alpha d\mu\right\} = \mathscr{P}\{X(t_i) \leq x_i, \ 1 \leq i \leq k\}.$$

For example, in the particular the case of moving maxima defined in (13), we have that (17) holds, where $U_t(g)(u) = g(t + u)$ is the simple translation in time and $f_0(u) = f(-u)$, for all $u \in \mathbb{R}$.

The representation in (17) is valid for a large class of stationary max–stable processes. In fact, as shown in Stoev [25], the above defined max–linear isometries are precisely the *pistons* of de Haan and Pickands [9]. Thus, by Theorem 6.1 in de Haan and Pickands [9], Relation (17) holds for all *continuous in probability* $\alpha$–Fréchet processes.

The following two results, established in Stoev [25], provide necessary and sufficient conditions for the ergodicity and mixing of the process $X$, respectively.

**Theorem 3.1 (Theorem 3.2 in [25]).** *Let $X$ be a measurable $\alpha$–Fréchet process, defined by (17). The process $X$ is ergodic, if and only if, for some (any) $p > 0$,*

$$\frac{1}{T} \int_0^T \|U_\tau g \wedge g\|_{L^\alpha(\mu)}^p d\tau \longrightarrow 0, \tag{18}$$

*as $T \to \infty$, for all $g \in F_U(f_0)$, where $a \wedge b = \min\{a, b\}$. Here*

$$F_U(f_0) := \overline{\text{V-span}}\{U_t(f_0),\ t \in \mathbb{R}\},$$

*is the set of all max–linear combinations of the $U_t(f_0)$'s, closed with respect to the metric $\rho_\alpha$ in (11).*

The corresponding necessary and sufficient condition for mixing is as follows

**Theorem 3.2 (Theorem 3.3 in [25]).** *Let $X$ be a measurable $\alpha$–Fréchet process, defined by (17). The process $X$ is mixing, if and only if,*

$$\|U_\tau h \wedge g\|_{L^\alpha(\mu)} \longrightarrow 0, \quad as\ \tau \to \infty, \tag{19}$$

*for all $g \in F_U^-(f_0) := \overline{\text{V-span}}\{U_t(f_0),\ t \leq 0\}$ and $h \in F_U^+(f_0) := \overline{\text{V-span}}\{U_t(f_0),\ t \geq 0\}$.*

Although these results provide complete characterization of the ergodic and/or mixing $\alpha$–Fréchet processes, they are hard to use in practice. This is because the conditions (18) and/or (19) should be verified for arbitrary elements $g$ and/or $h$ in the max–linear spaces $F_U(f_0)$ and/or $F_U^\pm(f_0)$. Fortunately, in the case of mixing, these conditions can be formulated simpler in terms of a natural *measure of dependence*. Namely, for any $\xi = \mathcal{J}_E f dM_\alpha$ and $\eta = \mathcal{J}_E g dM_\alpha$, $f, g \in L_+^\alpha(\mu)$ define

$$d(\xi, \eta) := \|\xi\|_\alpha^\alpha + \|\eta\|_\alpha^\alpha - \|\xi \vee \eta\|_\alpha^\alpha.$$

Observe that since $\|\xi\|_\alpha^\alpha = \int_E f^\alpha d\mu$ and $\|\eta\|_\alpha^\alpha = \int_E g^\alpha d\mu$, we have

$$d(\xi, \eta) = \int_E \left( f^\alpha + g^\alpha - f^\alpha \vee g^\alpha \right) d\mu \equiv \int_E f^\alpha \wedge g^\alpha d\mu. \tag{20}$$

Note that $d(\xi,\eta) = 0$ if and only if the random variables $\xi$ and $\eta$ are independent. This observation and the intuition about extremal integrals, suggest that the quantity $d(\xi,\eta)$ can be interpreted as a measure of dependence between $\xi$ and $\eta$. The following result established in Stoev [25] shows that $d(\xi,\eta)$ indeed plays such a role.

**Theorem 3.3 (Theorem 3.4 in [25]).** *Let X be a stationary and continuous in probability* $\alpha$ *–Fréchet process. The process X is mixing if and only if* $d_\alpha(X(\tau), X(0)) \rightarrow 0$, *as* $\tau \rightarrow \infty$.

**Remarks:**

1. Observe that by Theorem 3.2 and Relation (20), the condition $d(X_\tau, X_0) \rightarrow 0$, $\tau \rightarrow \infty$ is necessary for $X$ to be mixing. Surprisingly, Theorem 3.3 implies that this condition is also sufficient. In many situations it is easy to check whether the dependence coefficient $d(X_\tau, X_0)$ vanishes as the lag $\tau$ tends to infinity. The explicit knowledge of the max–linear isometries $U_t$ in (17) is not necessary.
2. The recent monograph of Dedecker *et al.* [10] provides many classes of remarkably flexible measures of dependence. To the best of my knowledge, these measures of dependence have not yet been studied in the context of max–stable processes. The knowledge of sharp inequalities involving these measures of dependence could lead to many interesting statistical results.

In the following section we will illustrate further the above results with concrete examples and applications.

# 4 Examples and Statistical Applications

## 4.1 Ergodic Properties of Some Max–Stable Processes

• *(Mixed Moving Maxima)* It is easy to show that all *moving maxima* and *mixed moving maxima processes* defined in (13) and (14) are mixing. Indeed, let

$$X(t) := \int_{\mathbb{R} \times V}^{e} f(t-u,v) M_\alpha(du,dv), \quad t \in \mathbb{R},$$

for some $f \in L_+^\alpha(du, v(dv))$, $\alpha > 0$ and observe that

$$d(X(t), X(0)) = \int_{\mathbb{R} \times V} f(t+u,v)^\alpha \wedge f(u,v)^\alpha du v(dv)$$

$$\leq 2 \int_{|u| \geq t/2} \left( \int_V f(u,v)^\alpha v(dv) \right) du. \tag{21}$$

The last inequality follows from the fact that for all $u \in \mathbb{R}$, and $t > 0$, either $|u| \geq t/2$ or $|t+u| \geq t/2$ and therefore,

$$f(t+u,v)^\alpha \wedge f(u,v)^\alpha \le f(t+u,v)^\alpha 1_{\{|t+u|\ge t/2\}} + f(u,v)^\alpha 1_{\{|u|\ge t/2\}}.$$

The inequality (21) and the integrability of $f^\alpha$ imply that $d(X(t),X(0)) \to 0$, as $t \to \infty$. This, in view of Theorem 3.3, implies that the mixed moving maxima process $X$ is mixing.

• *(Doubly Stochastic Processes)* As in the theory of sum–stable processes (see e.g. the monograph of Samorodnitsky and Taqqu [22]), we can associate a max–stable $\alpha$–Fréchet processes with any positive stochastic process $\xi = \{\xi(t)\}_{t\in T}$ with $\mathbb{E}\xi(t)^\alpha < \infty$ . Namely, suppose that $M_\alpha$ is a random sup–measure on a measure space $(E,\mathscr{E},\mu)$, where the control measure $\mu$ is now a *probability measure* (i.e. $\mu(E) = 1$). Any collection of spectral functions $\{f(t,u)\}_{t\in T} \subset L_+^\alpha(E,\mu(du))$ may be viewed as a stochastic process, defined on the probability space $(E,\mu)$. Conversely, a *non–negative* stochastic process $\xi = \{\xi(t)\}_{t\in T}$, defined on $(E,\mathscr{E},\mu)$, and such that $\mathbb{E}_\mu \xi(t)^\alpha = \int_E \xi(t,u)^\alpha \mu(du) < \infty$ may be used to define an $\alpha$–Fréchet process as follows:

$$X(t) := \int_E^e \xi(t,u) M_\alpha(du), \quad t \in T. \tag{22}$$

The $\alpha$–Fréchet process $X = \{X(t)\}_{t\in T}$ will be called *doubly stochastic*. Note that from the perspective of the random sup–measure $M_\alpha$, the integrands $\xi(t)$'s are *non–random* since they 'live' on a different probability space. The main benefit from this new way of defining a max–stable process $X$ is that one can use the properties of the stochastic process $\xi = \{\xi(t)\}_{t\in T}$ to establish the properties of the $\alpha$–Fréchet process $X$.

For example, let $\xi = \{\xi(t)\}_{t\in\mathbb{R}}$ be a strictly stationary, non–negative process on $(E,\mathscr{E},\mu)$ such that $\mathbb{E}_\mu \xi(t)^\alpha < \infty$. We then have that $X$ in (22) is also stationary. Indeed, for all $t_i \in \mathbb{R}$, $x_i > 0$, $1 \le i \le n$, and $h \in \mathbb{R}$, we have

$$\mathscr{P}\{X(t_i+h) \le x_i, \ 1 \le i \le n\} = \exp\left\{-\mathbb{E}_\mu\left(\bigvee_{1\le i\le n} \xi(t_i+h)/x_i\right)^\alpha\right\}$$

$$= \exp\left\{-\mathbb{E}_\mu\left(\bigvee_{1\le i\le n} \xi(t_i)/x_i\right)^\alpha\right\} = \mathscr{P}\{X(t_i) \le x_i, \ 1 \le i \le n\},$$

where in the second equality above we used the stationarity of $\xi$. Borrowing terminology from theory of sum–stable processes (see e.g. [3]), if the process $\xi$ is stationary, we call the $\alpha$–Fréchet process $X$ *doubly stationary*. The following result shows the perhaps surprising fact that if the process $\xi$ is *mixing*, then the doubly stationary process $X$ is *non–ergodic*.

**Proposition 0.1.** *Let* $X = \{X(t)\}_{t\in\mathbb{R}}$ *be a doubly stationary process defined as in* (22) *with non–zero* $\xi(t)$*'s. If the stationary process* $\xi = \{\xi(t)\}_{t\in\mathbb{R}}$ *is mixing, then* $X$ *is non–ergodic.*

*Proof.* Consider the quantity

$$d(X(t),X(0)) = \int_E \left(\xi(t,u) \wedge \xi(0,u)\right)^\alpha \mu(du) \equiv \mathbb{E}_\mu(\xi(t)^\alpha \wedge \xi(0)^\alpha).$$

We will show that $\liminf_{t\to\infty} d(X(t), X(0)) = c > 0$. This would then imply that the time–averages in Theorem 3.1 do not vanish, and hence $X$ is not ergodic.

Observe that since $\xi$ is mixing, for all Borel sets $A$, $B \subset \mathbb{R}$, we have

$$\mathscr{P}\{\xi(t) \in A, \, \xi(0) \in B\} \longrightarrow \mathscr{P}\{\xi(t) \in A\} \mathscr{P}\{\xi(0) \in B\}, \quad \text{as } t \to \infty.$$

Consider the intervals $A = B = (\varepsilon^{1/\alpha}, \infty)$, for some $\varepsilon > 0$, and note that the last relation is equivalent to

$$\mathscr{P}\{\xi(t)^\alpha \wedge \xi(0)^\alpha > \varepsilon\} \longrightarrow \mathscr{P}\{\xi(t)^\alpha > \varepsilon\} \mathscr{P}\{\xi(0)^\alpha > \varepsilon\}, \quad \text{as } t \to \infty. \quad (23)$$

Since the $\xi(t)$'s are not identically zero, there exists an $\varepsilon > 0$, such that $\mathscr{P}\{\xi(t)^\alpha > \varepsilon\} \equiv \mathscr{P}\{\xi(0)^\alpha > \varepsilon\} > 0$. Now, note that

$$\mathbb{E}(\xi(t)^\alpha \wedge \xi(0)^\alpha) \geq \varepsilon \mathscr{P}\{\xi(t)^\alpha \wedge \xi(0)^\alpha > \varepsilon\}.$$

This, in view of the convergence in (23) implies that

$$\liminf_{t\to\infty} \mathbb{E}(\xi(t)^\alpha \wedge \xi(0)^\alpha) > 0,$$

which as argued above, implies that the process $X$ is non–ergodic. $\square$

The above result suggests that most doubly stochastic $\alpha$–Fréchet processes are non–ergodic. This fact can be intuitively explained by the conceptual difference between the independence in the $\xi(t)$'s and the independence of their extremal integrals $X(t)$'s. Indeed, for $X(t)$ and $X(s)$ to be independent, one must have $\xi(t)\xi(s) = 0$, $\mu$–almost surely. The latter, unless the process $\xi$ is trivial, implies that $\xi(t)$ and $\xi(s)$ are dependent. The following example shows, however, that one can have *ergodic* and in fact *mixing* doubly stochastic processes. These processes will be stationary but *not* doubly stationary.

• (*Brown–Resnick Processes*) Let now $w = \{w(t)\}_{t\in\mathbb{R}}$ be a standard Brownian motion, defined on the probability space $(E, \mathscr{E}, \mu)$, i.e. $\{w(-t)\}_{t\geq 0}$ and $\{w(t)\}_{t\geq 0}$ are two independent standard Brownian motions. Introduce the non–negative process $\xi(t) := e^{w(t)/\alpha - |t|/2\alpha}$, $t \in \mathbb{R}$ and observe that $\mathbb{E}_\mu \xi(t)^\alpha = 1$. for all $t \in \mathbb{R}$.

The following doubly stochastic process $X = \{X(t)\}_{t\in\mathbb{R}}$ is said to be a *Brown–Resnick process*:

$$X(t) := \int_E^e \xi(t, u) M_\alpha(du) \equiv \int_E^e e^{w(t,u)/\alpha - |t|/2\alpha} M_\alpha(du), \quad t \in \mathbb{R}. \quad (24)$$

The max–stable process $\{\log X(t)\}_{t\geq 0}$ with $\alpha = 1$ and Gumbel marginals was first introduced by Brown and Resnick [2] as a limit involving extremes of Brownian motions. Surprisingly, the resulting max–stable process $X = \{X(t)\}_{t\in\mathbb{R}}$ is stationary. The one–sided stationarity of $X$ is easy to show, by using the fact that $\{w(t)\}_{t\geq 0}$ has stationary and independent increments (see e.g. [25]).

Recently, Kabluchko, Schlather and de Haan [15] studied general *doubly stochastic* processes of Brown–Resnick type. They established necessary and sufficient conditions for the stationarity of such max stable processes. The two–sided stationarity of the classical Brown–Resnick process $X$ above follows from their general results.

We now focus on the Brown–Resnick process in (24) and show that it is mixing. Indeed, the continuity in probability of $X$ follows from the $L^\alpha$–continuity of $\xi(t) = e^{w(t)/\alpha - |t|/2\alpha}$. Therefore, by Theorem 3.3, to prove that $X$ is mixing, it is enough to show that $d(X(t), X(0)) \to 0$, as $t \to \infty$. We have that, for all $t > 0$,

$$d(X(t), X(0)) = \mathbb{E}_\mu \left( e^{w(t) - t/2} \wedge e^{w(0)} \right) = \mathbb{E}_\mu \left( e^{\sqrt{t}Z - t/2} \wedge 1 \right),$$

where $Z$ is a standard Normal random variable under $\mu$. The last expectation is bounded above by:

$$\mathscr{P}\{Z > \sqrt{t}/2\} + \frac{1}{\sqrt{2\pi}} \int_{-\infty}^{\sqrt{t}/2} e^{\sqrt{t}z - t/2} e^{-z^2/2} dz =$$

$$\Phi(-\sqrt{t}/2) + \frac{1}{\sqrt{2\pi}} \int_{-\infty}^{\sqrt{t}/2} e^{-(z - \sqrt{t})^2/2} dz,$$

which equals $2\Phi(-\sqrt{t}/2)$, where $\Phi(t) = (2\pi)^{-1/2} \int_{-\infty}^t e^{-x^2/2} dx$. Therefore,

$$d(X(t), X(0)) \le 2\Phi(-\sqrt{t}/2) \le \frac{2}{\sqrt{2\pi}} e^{-t^2/2} \longrightarrow 0, \quad \text{as } t \to \infty.$$

This implies that the Brown–Resnick process $X$ is mixing.

In [25], the ergodicity of more general Brown–Resnick type processes was established where the process $w$ in (24) is replaced by certain infinitely divisible Lévy processes. It would be interesting to define and study other classes of doubly stochastic processes by using different types of integrands.

## 4.2 Estimation of the Extremal Index

The extremal index is an important statistical quantity that can be used to measure the asymptotic *dependence* of stationary sequences. Here, we will briefly review the definition of the extremal index and discuss some estimators for the special case of max–stable time series.

Let $Y = \{Y_k\}_{k \in \mathbb{Z}}$ be strictly stationary time series, which is not necessarily max–stable. Associate with $Y$ a sequence of independent and identically distributed variables $Y^* = \{Y_k^*\}_{k \in \mathbb{R}}$, with the *same marginal* distribution as the $Y_k$'s. Consider the running maxima $M_n := \max_{1 \le j \le n} Y_j$ and $M_n^* := \max_{1 \le j \le n} Y_j^*$ and suppose that

$$\mathscr{P}\left\{ \frac{1}{a_n} M_n^* - b_n \le x \right\} \xrightarrow{w} G(x), \quad \text{as } n \to \infty, \tag{25}$$

where $G$ is one of the three extreme value distribution functions. If the last convergence holds, for suitable normalization and centering constants $u_n$ and $b_n$, then we say that the distribution of the $Y_k$'s belongs to the maximum domain of attraction of $G$ (for more details, see e.g. [18]).

**Definition 0.3.** Suppose that (25) holds for the maxima of independent $Y_k^*$'s. We say that the time series $Y$ has an *extremal index* $\theta$, if

$$\mathscr{P}\left\{\frac{1}{a_n}M_n - b_n \leq x\right\} \xrightarrow{w} G^\theta(x), \quad \text{as } n \to \infty, \tag{26}$$

where the $a_n$'s and $b_n$'s are as in (25).

It turns out that if the time series $Y = \{Y_k\}_{k\in\mathbb{Z}}$ has an extremal index $\theta$, then it necessarily follows that

$$0 \leq \theta \leq 1.$$

Observe that if the $Y_k$'s are independent and belong to the maximum domain of attraction of an extreme value distribution, then trivially, $Y$ has extremal index $\theta = 1$. The converse however, is not true, that is, $\theta = 1$ does not imply the independence of the $Y_k$'s, in general. It is important that the centering and normalization sequences in (25) and (26) be the same. For more details, see the monograph of Leadbetter, Lindgren and Rootzén [16].

A number of statistics have been proposed for the estimation of the extremal index (see e.g. [23], [28], [12]). Here, our goal is to merely illustrate the use of some new estimation techniques for the extremal index in the special case of max–stable $\alpha$–Fréchet processes. We propose a method to construct asymptotically consistent upper bounds for the extremal index, if it exists. The detailed analysis of these methods for the case of general time series is beyond the scope of this work.

Suppose now that $Y = \{Y_k\}_{k\in\mathbb{Z}}$ is a stationary, max–stable time series with the following extremal integral representation

$$Y_k = \int_E^e f_k(u)M_\alpha(du), \quad k \in \mathbb{Z}, \tag{27}$$

The extremal index of $Y = \{Y_k\}_{k\in\mathbb{Z}}$, if it exists, can be expressed simply as follows. By the max–linearity of the extremal integrals, we have with $M_n$ as in (26), that:

$$\mathscr{P}\left\{\frac{1}{n^{1/\alpha}}M_n \leq x\right\} = \exp\left\{-\frac{1}{n}\int_E^e \left(\bigvee_{1\leq k\leq n} f_k^\alpha\right)d\mu x^{-\alpha}\right\}.$$

On the other hand, by the independence of the $Y_k^*$'s, we have

$$\mathscr{P}\left\{\frac{1}{n^{1/\alpha}}M_n^* \leq x\right\} = \exp\{-\|Y_1\|_\alpha^\alpha x^{-\alpha}\},$$

where $\|Y_1\|_\alpha^\alpha = \int_E f_1^\alpha d\mu$.

Thus, the extremal index of $Y$ exists and equals $\theta$, if and only if, the following limit exists:

$$\theta := \frac{1}{\|Y_1\|_\alpha^\alpha} \lim_{n\to\infty} \frac{1}{n} \int_E \left( \bigvee_{1\le k\le n} f_k^\alpha \right) d\mu. \tag{28}$$

This fact suggests simple and intuitive ways of expressing the extremal index of $Y$. To do so, let $r \in \mathbb{N}$ and consider the time series $Y(r) := \{Y_k(r)\}_{k\in\mathbb{Z}}$ of non–overlapping block–maxima of size $r$:

$$Y_k(r) := \max_{1\le i\le r} Y_{i+(k-1)r}, \quad k \in \mathbb{Z}.$$

Observe that $Y = Y(1)$ is the original time series. We shall denote the extremal index of $Y(r)$ as $\theta(r)$, when it exists. The following result yields a simple relationship between the $\theta(r)$'s.

**Proposition 0.2.** *Let $Y = \{Y_k\}_{k\in\mathbb{Z}}$ be as in (27). If $Y$ has a positive extremal index $\theta = \theta(1)$, then the time series $Y(r) = \{Y_k(r)\}_{k\in\mathbb{Z}}$ also has an extremal index $\theta(r)$ equal to:*

$$\theta(r) = \frac{1}{\theta_r(1)}\theta(1),$$

*where*

$$\theta_r(1) = \frac{\|Y_1\|_\alpha^{-\alpha}}{r} \int_E \left( \bigvee_{1\le k\le r} f_k^\alpha \right) d\mu, \quad r \in \mathbb{N}. \tag{29}$$

*Moreover, for all $r \in \mathbb{N}$, we have $\theta_r(1) \in (0,1]$ and hence*

$$\theta(1) \le \theta(r) \quad and \quad \theta(1) \le \theta_r(1). \tag{30}$$

We will see in the sequel that, for fixed $r$, the quantity $\theta_r(1)$ can be consistently estimated from the data, provided that the underlying time series $Y$ is ergodic.

*Proof (Proposition 0.2).* In view of (28), we have that

$$\theta(r) = \|Y_1 \vee \cdots \vee Y_r\|_\alpha^{-\alpha} \lim_{n\to\infty} \frac{1}{n} \int_E \left( \bigvee_{1\le k\le n} \bigvee_{1\le i\le r} f_{i+(k-1)r}^\alpha \right) d\mu$$

$$= r\|Y_1 \vee \cdots \vee Y_r\|_\alpha^{-\alpha} \lim_{n\to\infty} \frac{1}{rn} \int_E \left( \bigvee_{1\le j\le nr} f_j^\alpha \right) d\mu, \tag{31}$$

where in the last relation we multiplied and divided by the constant $r$.
By assumption, the limit in the right–hand side of (31) exists and equals $\|Y_1\|_\alpha^\alpha \theta(1)$, which implies that the time series $Y(r)$ has an extremal index $\theta(r)$. This, fact since

$$\|Y_1 \vee \cdots \vee Y_r\|_\alpha^\alpha = \int_E \left( \bigvee_{1\le k\le r} f_k^\alpha \right) d\mu$$

and in view of (29) also implies that $\theta(r) = \theta(1)/\theta_r(1)$.

Finally, the inequalities in (30) follow from the fact that

$$\frac{1}{r} \int_E \Big( \bigvee_{1 \leq k \leq r} f_k^\alpha \Big) d\mu \leq \frac{1}{r} \sum_{k=1}^r \int_E f_k^\alpha d\mu = \|Y_1\|_\alpha^\alpha,$$

which yields that $0 < \theta_r(1) \leq 1$. $\square$

We now focus on the estimation of the parameter $\theta_r(1)$ for a given fixed value of $r$, from observations $\{Y_k, \ 1 \leq k \leq n\}$ of the time series $Y$. Suppose that the time series $Y$ is ergodic. Note that $\mathbb{E}Y_k^p = \Gamma(1 - p/\alpha)\|Y_1\|_\alpha^p$ is finite, for all $p < \alpha$. Therefore, the ergodicity implies that

$$\widehat{m}_p(1) := \frac{1}{n} \sum_{k=1}^n Y_k^p \xrightarrow{a.s.} \Gamma(1 - p/\alpha)\|Y_1\|_\alpha^p, \tag{32}$$

as $n \to \infty$. For the block–maxima time series, we also have that

$$\widehat{m}_p(r) := \frac{1}{[n/r]} \sum_{k=1}^{[n/r]} Y_k(r)^p \xrightarrow{a.s.} \Gamma(1 - p/\alpha)\|Y_1 \vee \cdots \vee Y_r\|_\alpha^p. \tag{33}$$

Note that here we have only $[n/r]$ observations from the block–maxima time series $\{Y_k(r), \ 1 \leq k \leq [n/r]\}$ available from the original data set.

Relation (29) and the convergences in (32) and (33) suggest the following estimator for the parameter $\theta_r(1)$:

$$\widehat{\theta}_r(1; p, n) := \frac{1}{r} \Big( \widehat{m}_p(r)/\widehat{m}_p(1) \Big)^{\alpha/p}. \tag{34}$$

**Proposition 0.3.** *Suppose that $\{Y_k, \ 1 \leq k \leq n\}$ is a sample from an ergodic stationary $\alpha$–Fréchet time series. Then, for all $p < \alpha$, we have*

$$\widehat{\theta}_r(1; p, n) \xrightarrow{a.s.} \theta_r(1), \quad as \ n \to \infty.$$

This result shows the strong consistency of the estimator in (34). The proof of this proposition is an immediate consequence from Relations (32) and (33).

Note that one can use also *overlapping* block–maxima to estimate the quantity $\|Y_1(r)\|_\alpha^\alpha = \|Y_1 \vee \cdots \vee Y_r\|_\alpha^\alpha$. Indeed, for an ergodic time series $Y$, we also have

$$\widehat{m}_{p,ovlp}(1; p, n) :=$$

$$\frac{1}{n-r+1} \sum_{k=1}^{n-r+1} (Y_k \vee Y_{k+1} \vee \cdots \vee Y_{k+r-1})^p \xrightarrow{a.s.} \Gamma(1 - p/\alpha)\|Y_1(r)\|_\alpha^p,$$

as $n \to \infty$. This suggests another flavor of an estimator for $\theta_r(1)$:

$$\widehat{\theta}_{r,ovlp}(1; p, n) := \frac{1}{r} \Big( \widehat{m}_{p,ovlp}(r)/\widehat{m}_p(1) \Big)^{\alpha/p}.$$

Clearly, as for the estimator $\widehat{\theta}_r$, we also have *strong consistency*:

$$\widehat{\theta}_{r,\text{ovlp}}(1;p,n) \xrightarrow{a.s.} \theta_r(1), \quad \text{as } n \to \infty.$$

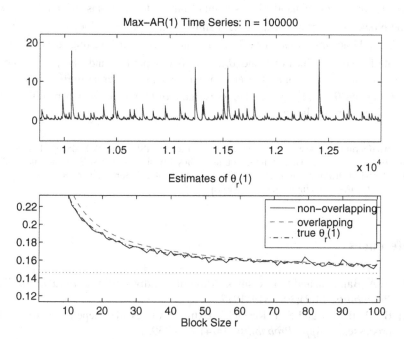

**Fig. 1** *Top panel*: Simulated max–autoregressive $\alpha$–Fréchet time series $Y_k$, $1 \leq k \leq n$ defined as in (35) with $n = 100\,000$, $\phi = 0.9$ and $\alpha = 1.5$. The theoretical value of $\theta = \theta(1)$ equals $(1 - \phi^\alpha) = 0.1462$. *Bottom panel*: Estimates $\widehat{\theta}_r(1)$ (solid line) and $\widehat{\theta}_{r,\text{ovlp}}(1)$ (dashed line) as a function of the block–size $r$. The true value of $\theta(1)$ is indicated by the dotted line.

Figure 1 illustrates the performance of the estimators $\widehat{\theta}_r(1)$ and $\widehat{\theta}_{r,\text{ovlp}}(1)$ over a simulated max–autoregressive time series. The time series $Y$ is defined as:

$$Y_k = \phi Y_{k-1} \vee (1 - \phi)Z_k = (1 - \phi) \bigvee_{j=0}^{\infty} \phi^j Z_{k-j}, \tag{35}$$

where $\phi \in [0,1]$, and where the $Z_j$'s are independent standard $\alpha$–Fréchet variables. We used parameter values $\alpha = 1.5$, $\phi = 0.9$ and $p = 0.6$.

One can show that $\|Y_1\|_\alpha^\alpha = (1 - \phi)^\alpha/(1 - \phi^\alpha)$, and that

$$Y_1 \vee \cdots \vee Y_r = Y_1 \vee (1 - \phi)\big(\max_{2 \leq k \leq r} Z_k\big).$$

Thus, for $\theta_r(1)$ and $\theta(1)$ we obtain:

$$\theta_r(1) = (1 - \phi^\alpha)\frac{(r-1)}{r} + \frac{1}{r} \quad \text{and} \quad \theta(1) = \lim_{r\to\infty} \theta_r(1) = (1 - \phi^\alpha).$$

The true value of $\theta_r(1)$, indicated by the dot–dashed line is nearly covered by the solid line, indicating the realizations of $\widehat{\theta}_r(1)$. The 'overlapping blocks' estimator $\widehat{\theta}_{r,\text{ovlp}}(1)$, on the other hand, shows a small but systematic bias, which decreases as the block–size $r$ grows. This limited simulation experiment indicates that $\widehat{\theta}_r(1)$ and $\widehat{\theta}_{r,\text{ovlp}}(1)$ accurately estimate $\theta_r(1)$. Furthermore, in this setting, $\widehat{\theta}_r(1)$ is more accurate for small values of $r$ and $\widehat{\theta}_{r,\text{ovlp}}(1)$ is competitive and likely to be more accurate for large values of $r$. Note also that since $\theta_r(1)$ converges to $\theta(1)$, as $r \to \infty$, both $\widehat{\theta}_r(1)$ and $\widehat{\theta}_{r,\text{ovlp}}(1)$ can be used to estimate $\theta(1)$ in practice, when sufficiently large values of $r$ are chosen.

**Acknowledgements** The author was partially supported by NSF grant DMS–0806094 at the University of Michigan. We thank Philippe Soulier, Gilles Teyssière and INSEE for their hospitality during the STATDEP meeting in Paris, 2008. We also thank an anonymous referee for insightful comments that helped us improve the manuscript.

# References

[1] A. A. Balkema and S. I. Resnick. Max-infinite divisibility. *Journal of Applied Probability*, 14(2):309–319, 1977.

[2] B. M. Brown and S. I. Resnick. Extreme values of independent stochastic processes. *J. Appl. Probability*, 14(4):732–739, 1977.

[3] S. Cambanis, C. D. Hardin, Jr., and A. Weron. Ergodic properties of stationary stable processes. Technical Report 59, Center for Stochastic Processes at the University of North Carolina, Chapel Hill, 1984.

[4] L. de Haan. A characterization of multidimensional extreme-value distributions. *Sankhyā (Statistics). The Indian Journal of Statistics. Series A*, 40(1):85–88, 1978.

[5] L. de Haan. A spectral representation for max–stable processes. *Annals of Probability*, 12(4):1194–1204, 1984.

[6] L. de Haan and A. Ferreira. *Extreme value theory*. Springer Series in Operations Research and Financial Engineering. Springer, New York, 2006. An introduction.

[7] L. de Haan and T. Lin. On convergence toward an extreme value distribution in $C[0,1]$. *The Annals of Probability*, 29(1):467–483, 2001.

[8] L. de Haan and T. Lin. Weak consistency of extreme value estimators in $C[0,1]$. *The Annals of Statistics*, 31(6):1996–2012, 2003.

[9] L. de Haan and J. Pickands III. Stationary min–stable stochastic processes. *Probability Theory and Related Fields*, 72(4):477–492, 1986.

[10] J. Dedecker, P. Doukhan, G. Lang, J. R. León R., S. Louhichi, and C. Prieur. *Weak dependence: with examples and applications*, volume 190 of *Lecture Notes in Statistics*. Springer, New York, 2007.

[11] P. Deheuvels. Point processes and multivariate extreme values. *Journal of Multivariate Analysis*, 13(2):257–272, 1983.

[12] C. A. T. Ferro and J. Segers. Inference for clusters of extreme values. *J. R. Stat. Soc. Ser. B Stat. Methodol.*, 65(2):545–556, 2003.

[13] E. Giné, M. Hahn, and P. Vatan. Max–infinitely divisible and max–stable sample continuous processes. *Probability Theory and Related Fields*, 87(2):139–165, 1990.

[14] H. Hult and F. Lindskog. Extremal behavior of regularly varying stochastic processes. *Stochastic Processes and their Applications*, 115(2):249–274, 2005.

[15] Z. Kabluchko, M. Schlather, and L. de Haan. Stationary max–stable fields associated to negative definite functions. Available from http://axiv.org: *arXiv*:0806.2780v1, 2008.

[16] M. R. Leadbetter, G. Lindgren, and H. Rootzén. *Extremes and Related Properties of Random Sequences and Processes*. Springer-Verlag, New York, 1983.

[17] T. Norberg. *Random sets and capacities, with applications to extreme value theory*. University of Göteborg, Göteborg, 1985.

[18] S. I. Resnick. *Extreme Values, Regular Variation and Point Processes*. Springer-Verlag, New York, 1987.

[19] S. I. Resnick. *Heavy-tail phenomena*. Springer Series in Operations Research and Financial Engineering. Springer, New York, 2007. Probabilistic and statistical modeling.

[20] S. I. Resnick and R. Roy. Random USC functions, max-stable processes and continuous choice. *The Annals of Applied Probability*, 1(2):267–292, 1991.

[21] Y. A. Rozanov. *Stationary Random Processes*. Holden-Day, San Francisco, 1967.

[22] G. Samorodnitsky and M. S. Taqqu. *Stable Non-Gaussian Processes: Stochastic Models with Infinite Variance*. Chapman and Hall, New York, London, 1994.

[23] R. L. Smith and I. Weissman. Estimating the extremal index. *J. Roy. Statist. Soc. Ser. B*, 56(3):515–528, 1994.

[24] S. Stoev and M. S. Taqqu. Extremal stochastic integrals: a parallel between max–stable processes and $\alpha$–stable processes. *Extremes*, 8:237–266, 2005.

[25] S. A. Stoev. On the ergodicity and mixing of max-stable processes. *Stochastic Process. Appl.*, 118(9):1679–1705, 2008.

[26] W. Vervaat. Stationary self-similar extremal processes and random semi-continuous functions. In *Dependence in probability and statistics (Oberwolfach, 1985)*, volume 11 of *Progr. Probab. Statist.*, pages 457–473. Birkhäuser Boston, Boston, MA, 1986.

[27] K. S. Weintraub. Sample and ergodic properties of some min–stable processes. *The Annals of Probability*, 19(2):706–723, 1991.

[28] I. Weissman and S. Y. Novak. On blocks and runs estimators of the extremal index. *J. Statist. Plann. Inference*, 66(2):281–288, 1998.
[29] Z. Zhang and R. Smith. The behavior of multivariate maxima of moving maxima processes. *Journal of Applied Probability*, 41(4):1113–1123, 2004.

# Best attainable rates of convergence for the estimation of the memory parameter

Philippe Soulier

**Abstract** The purpose of this note is to prove a lower bound for the estimation of the memory parameter of a stationary long memory process. The memory parameter is defined here as the index of regular variation of the spectral density at 0. The rates of convergence obtained in the literature assume second order regular variation of the spectral density at zero. In this note, we do not make this assumption, and show that the rates of convergence in this case can be extremely slow. We prove that the log-periodogram regression (GPH) estimator achieves the optimal rate of convergence for Gaussian long memory processes

## 1 Introduction

Let $\{X_n\}$ be a weakly stationary process with autocovariance function $\gamma$. Its spectral density $f$, when it exists, is an even nonnegative measurable function such that

$$\gamma(k) = \int_{-\pi}^{\pi} f(x)\mathrm{e}^{\mathrm{i}kx}\, \mathrm{d}x\,.$$

Long memory of the weakly stationary process $\{X_t\}$ means at least that the auto-covariance function is not absolutely summable. This definition is too weak to be useful. It can be strengthen in several ways. We will assume here that the spectral density is regularly varying at zero with index $-\alpha \in (-1,1)$, i.e. it can be expressed for $x \geq 0$ as

$$f(x) = x^{-\alpha}L(x)\,,$$

where the function $L$ is slowly varying at zero, which means that for all $t > 0$,

Philippe Soulier
Université Paris Ouest-Nanterre, 200 avenue de la République,
92000 Nanterre cedex, France, e-mail: philippe.soulier@u-paris10.fr

P. Doukhan et al. (eds.), *Dependence in Probability and Statistics*,
Lecture Notes in Statistics 200, DOI 10.1007/978-3-642-14104-1_3,
© Springer-Verlag Berlin Heidelberg 2010

$$\lim_{x \to 0} \frac{L(tx)}{L(x)} = 1 .$$

Then the autocovariance function is regularly varying at infinity with index $\alpha -$ 1 and non absolutely summable for $\alpha > 0$. The main statistical problem for long memory processes is the estimation of the memory parameter $\alpha$. This problem has been exhaustively studied for the most familiar long memory models: the fractional Gaussian noise and the ARFIMA($p,d,q$) process. The most popular estimators are the GPH estimator and the GSE estimator, first introduced respectively by [4] and [10]. Rigorous theoretical results for these estimators were obtained by [12, 13], under an assumption of second order regular variation at 0, which roughly means that there exists $C, \rho > 0$ such that

$$f(x) = Cx^{-\alpha}\{1 + O(x^{\rho})\} .$$

Under this assumption, [5] proved that the optimal rate of convergence of an estimator based on a sample of size $n$ is of order $n^{2\rho/(2\rho+1)}$.

The methodology to prove these results is inspired from similar results in tail index estimation. If $F$ is a probability distribution function on $(-\infty, \infty)$ which is second order regularly varying at infinity, i.e. such that

$$\bar{F}(x) = Cx^{-\alpha}\{1 + O(x^{-\alpha\rho})\}$$

as $x \to \infty$, then [6] proved that the best attainable rate of convergence of an estimator of the tail index $\alpha$ based on $n$ i.i.d. observations drawn from the distribution $F$ is of order $n^{2\rho/(2\rho+1)}$. In this context, [3] first considered the case where the survival function $\bar{F}$ is regularly varying at infinity, but not necessarily second order regularly varying. He introduced very general classes of slowly varying functions for which optimal rates of convergence of estimators of the tail index can be computed. The main finding was that the rate of convergence can be extremely slow in such a case.

In the literature on estimating the memory parameter, the possibility that the spectral density is not second order regularly varying has not yet been considered. Since this has severe consequences on the estimations procedures, it seems that this problem should be investigated. In this note, we parallel the methodology developed by [3] to deal with such regularly varying functions. Not surprisingly, we find the same result, which show that the absence of second order regular variation of the spectral density has the same drastic consequences.

The rest of the paper is organized as follows. In Section 2, we define the classes of slowly varying functions that will be considered and prove a lower bound for the rate of convergence of the memory parameter. This rate is proved to be optimal in Section 3. An illustration of the practical difficulty to choose the bandwidth parameter is given in Section 4. Technical lemmas are deferred to Section 5.

## 2 Lower bound

In order to derive precise rates of convergence, it is necessary to restrict attention to the class of slowly varying functions referred to by [3] as *normalized*. This class is also referred to as the Zygmund class. Cf. [2, Section 1.5.3]

**Definition 0.1.** Let $\eta^*$ be a non decreasing function on $[0, \pi]$, regularly varying at zero with index $\rho \geq 0$ and such that $\lim_{x \to 0} \eta^*(x) = 0$. Let $SV(\eta^*)$ be the class of even measurable functions $L$ defined on $[-\pi, \pi]$ which can be expressed for $x \geq 0$ as

$$L(x) = L(\pi) \exp\left\{ -\int_x^\pi \frac{\eta(s)}{s} \, ds \right\},$$

for some measurable function $\eta$ such that $|\eta| \leq \eta^*$.

This representation implies that $L$ has locally bounded variations and $\eta(s) = sL'(s)/L(s)$. Usual slowly varying functions, such as power of logarithms, iterated logarithms are included in this class, and it easy to find the corresponding $\eta$ function. Examples are given below. We can now state our main result.

**Theorem 2.1.** *Let $\eta^*$ be a non decreasing function on $[0, \pi]$, regularly varying at $0$ with index $\rho \geq 0$ and such that $\lim_{x \to 0} \eta^*(x) = 0$. Let $t_n$ be a sequence satisfying*

$$\lim_{n \to \infty} \eta^*(t_n)(nt_n)^{1/2} = 1. \tag{1}$$

*Then, if $\rho > 0$,*

$$\liminf_{n \to \infty} \inf_{\widehat{\alpha}_n} \sup_{L \in SV(\eta^*)} \sup_{\alpha \in (-1,1)} \mathbb{E}_{\alpha,L}[\eta^*(t_n)^{-1}|\widehat{\alpha}_n - \alpha|] > 0, \tag{2}$$

*and if $\rho = 0$*

$$\liminf_{n \to \infty} \inf_{\widehat{\alpha}_n} \sup_{L \in SV(\eta^*)} \sup_{\alpha \in (-1,1)} \mathbb{E}_{\alpha,L}[\eta^*(t_n)^{-1}|\widehat{\alpha}_n - \alpha|] \geq 1, \tag{3}$$

*where $\mathbb{P}_{\alpha,L}$ denotes the distribution of any second order stationary process with spectral density $x^{-\alpha}L(x)$ and the infimum $\inf_{\widehat{\alpha}_n}$ is taken on all estimators of $\alpha$ based on $n$ observations of the process.*

*Example 2.1.* Define $\eta^*(s) = Cs^\beta$ for some $\beta > 0$ and $C > 0$. Then any function $L \in SV(\eta^*)$ satisfies $L(x) = L(0) + O(x^\beta)$, and we recover the case considered by [5]. The lower bound for the rate of convergence is $n^{\beta/(2\beta+1)}$.

*Example 2.2.* For $\rho > 0$, define $\eta^*(s) = \rho/\log(1/s)$, then

$$\exp\left\{ \int_x^{1/e} \frac{\eta^*(s)}{s} \, ds \right\} = \exp\{\rho \log\log(1/x)\} = \log^\rho(1/x).$$

A suitable sequence $t_n$ must satisfy $\rho^2/\log^2(t_n) \approx nt_n$. One can for instance choose $t_n = \log^2(n)/(n\rho^2)$, which yields $\eta^*(t_n) = \rho/\log(n)\{1 + o(1)\}$. Note that $\eta(s) =$

$\rho/\log(s)$ belongs to $SV(\eta^*)$, and the corresponding slowly varying function is $\log^{-\rho}(1/x)$. Hence, the rate of convergence is not affected by the fact that the slowly varying function vanishes or is infinite at 0.

*Example 2.3.* The function $L(x) = \log\log(1/x)$ is in the class $SV(\eta^*)$ with $\eta^*(x) = \{\log(1/x)\log\log(1/x)\}^{-1}$. In that case, the optimal rate of convergence is $\log(n)$ $\log\log(n)$. Even though the slowly varying function affecting the spectral density at zero diverges very weakly, the rate of convergence of any estimator of the memory parameter is dramatically slow.

**Proof of Theorem 2.1** Let $\ell > 0$, $t_n$ be a sequence that satisfies the assumption of Theorem 2.1, and define $\alpha_n = \eta^*(\ell t_n)$ and

$$\eta_n(s) = \begin{cases} 0 & \text{if } 0 \le s \le \ell t_n, \\ \alpha_n & \text{if } \ell t_n < s \le \pi, \end{cases}$$

$$L_n(x) = \pi^{\alpha_n} \exp\left\{-\int_x^\pi \eta_n(s)\,ds\right\}.$$

Since $\eta^*$ is assumed non decreasing, it is clear that $L_n \in SV(\eta^*)$. Define now $f_n^-(x) = x^{-\alpha_n} L_n(x)$ and $f_n^+ = (f_n^-)^{-1}$. $f_n^-$ can be written as

$$f_n^-(x) = \begin{cases} (\ell t_n/x)^{\alpha_n} & \text{if } 0 < x \le \ell t_n, \\ 1 & \text{if } \ell t_n < x \le \pi. \end{cases}$$

Straightforward computations yield

$$\int_0^\pi \{f_n^-(x) - f_n^+(x)\}^2 \, dx = 8\ell t_n \alpha_n^2 (1 + O(\alpha_n^2)) = 8\ell n^{-1}(1 + o(1)). \tag{4}$$

The last equality holds by definition of the sequence $t_n$. Let $\mathbb{P}_n^-$ and $\mathbb{P}_n^+$ denote the distribution of a $n$-sample of a stationary Gaussian processes with spectral densities $f_n^-$ et $f_n^+$ respectively, $\mathbb{E}_n^-$ and $\mathbb{E}_n^+$ the expectation with respect to these probabilities, $\frac{d\mathbb{P}_n^+}{d\mathbb{P}_n^-}$ the likelihood ratio and $A_n = \{\frac{d\mathbb{P}_n^+}{d\mathbb{P}_n^-} \ge \tau\}$ for some real $\tau \in (0,1)$. Then, for any estimator $\widehat{\alpha}_n$, based on the observation $(X_1, \ldots, X_n)$,

$$\sup_{\alpha,L} \mathbb{E}_{\alpha,L}[|\widehat{\alpha}_n - \alpha|] \ge \frac{1}{2}\left(\mathbb{E}_n^+[|\widehat{\alpha}_n - \alpha_n|] + \mathbb{E}_n^-[|\widehat{\alpha}_n + \alpha_n|]\right)$$

$$\ge \frac{1}{2}\mathbb{E}_n^-\left[\mathbb{1}_{A_n}|\widehat{\alpha}_n + \alpha_n| + \frac{d\mathbb{P}_n^+}{d\mathbb{P}_n^-}\mathbb{1}_{A_n}|\widehat{\alpha}_n - \alpha_n|\right]$$

$$\ge \frac{1}{2}\mathbb{E}_n^-[\{|\widehat{\alpha}_n + \alpha_n| + \tau|\widehat{\alpha}_n - \alpha_n|\}\mathbb{1}_{A_n}] \ge \tau\alpha_n\mathbb{P}_n^-(A_n).$$

Denote $\varepsilon = \log(1/\tau)$ and $\Lambda_n = \log(d\mathbb{P}_n^+/d\mathbb{P}_n^-)$. Then $\mathbb{P}_n^-(A_n) = 1 - \mathbb{P}_n^-(\Lambda_n \le -\varepsilon)$. Applying (4) and [5, Lemma 2], we obtain that there exist constants $C_1$ and $C_2$ such that

$$\mathbb{E}_n^-[\Lambda_n] \le C_1\ell, \quad \mathbb{E}_n^-[(\Lambda_n - m_n)^2] \le C_2\ell.$$

This yields, for any $\eta > 0$ and small enough $\ell$,

$$\mathbb{P}_n^-(A_n) \geq 1 - \varepsilon^{-2}\mathbb{E}[\Lambda_n^2] \geq 1 - C\ell\varepsilon^{-2} \geq 1 - \eta \,.$$

Thus, for any $\eta, \tau \in (0,1)$, and sufficiently small $\ell$, we have

$$\liminf_{n\to\infty} \inf_{L\in SV(\eta^*)} \inf_{\alpha\in(-1,1)} \mathbb{E}_{\alpha,L}[\eta^*(t_n)^{-1}|\widehat{\alpha}_n - \alpha|]$$

$$\geq \tau(1-\eta)\lim_{n\to\infty} \frac{\eta^*(\ell t_n)}{\eta^*(t_n)} = \tau(1-\eta)\ell^\rho \,.$$

This proves (2) and (3).                                                                        $\diamondsuit$

# 3 Upper bound

In the case $\eta^*(x) = Cx^\rho$ with $\rho > 0$, [5] have shown that the lower bound (2) is attainable. The extension of their result to the case where $\eta^*$ is regularly varying with index $\rho > 0$ (for example to functions of the type $x^\beta \log(x)$) is straightforward. We will restrict our study to the case $\rho = 0$, and will show that the lower bound (3) is asymptotically sharp, i.e. there exist estimators that are rate optimal up to the exact constant.

Define the discrete Fourier transform and the periodogram ordinates of a process $X$ based on a sample $X_1,\dots,X_n$, evaluated at the Fourier frequencies $x_j = 2j\pi/n$, $j = 1,\dots,n$, respectively by

$$d_{X,j} = (2\pi n)^{-1/2} \sum_{t=1}^n X_t e^{-itx_j}, \quad \text{and } I_{X,j} = |d_{X,j}|^2.$$

The frequency domain estimates of the memory parameter $\alpha$ are based on the following heuristic approximation: the renormalized periodogram ordinate $I_{X,j}/f(x_j)$, $1 \leq j \leq n/2$ are approximately i.i.d. standard exponential random variables. Although this is not true, the methods and conclusion drawn from these heuristics can be rigorously justified. In particular, the Geweke and Porter-Hudak (GPH) and Gaussian semiparametric estimator have been respectively proposed by [4] and [10], and a theory for them was obtained by [13, 12] in the case where the spectral density is second order regularly varying at 0.

The GPH estimator is based on an ordinary least square regression of $\log(I_{X,k})$ on $\log(k)$ for $k = 1,\dots,m$, where $m$ is a bandwidth parameter:

$$(\widehat{\alpha}(m),\widehat{C}) = \arg\min_{\alpha,C} \sum_{k=1}^m \left\{\log(I_{X,k}) - C + \alpha\log(k)\right\}^2.$$

The GPH estimator has an explicit expression as a weighted sum of log-periodogram ordinates:

$$\widehat{\alpha}(m) = -s_m^{-2} \sum_{k=1}^{m} v_{m,k} \log(I_{X,k}),$$

with $v_{m,k} = \log(k) - m^{-1} \sum_{j=1}^{m} \log(j)$ and $s_m^2 = \sum_{k=1}^{m} v_{m,k}^2 = m\{1 + o(1)\}$.

**Theorem 3.1.** *Let $\eta^*$ be a non decreasing slowly varying function such that $\lim_{x \to 0} \eta^*(x) = 0$. Let $\mathbb{E}_{\alpha,L}$ denote the expectation with respect to the distribution of a Gaussian process with spectral density $x^{-\alpha} L(x)$. Let $t_n$ be a sequence that satisfies (1) and let $m$ be a non decreasing sequence of integers such that*

$$\lim_{n \to \infty} m^{1/2} \eta^*(t_n) = \infty \ and \ \lim_{n \to \infty} \frac{\eta^*(t_n)}{\eta^*(m/n)} = 1 . \tag{5}$$

*Assume also that the sequence $m$ can be chosen in such a way that*

$$\lim_{n \to \infty} \frac{\log(m) \int_{m/n}^{\pi} s^{-1} \eta^*(s) \, ds}{m \eta^*(m/n)} = 0 . \tag{6}$$

*Then, for any $\delta \in (0,1)$,*

$$\limsup_{n \to \infty} \sup_{|\alpha| \leq \delta} \sup_{L \in SV(\eta^*)} \eta^*(t_n)^{-2} \mathbb{E}_{\alpha,L}[(\widehat{\alpha}(m) - \alpha)^2] \leq 1. \tag{7}$$

*Remark 0.1.* Since $\eta^*$ is slowly varying, it is always possible to choose the sequence $m$ in such a way that (5) holds. Condition (6) ensures that the bias of the estimator is of the right order. It is very easily checked and holds for all the examples of usual slowly varying function $\eta*$, but we have not been able to prove that it always holds.

Since the quadratic risk is greater than the $L^1$ risk, we obtain the following corollary.

**Corollary 3.1.** *Let $\delta \in (0,1)$ and $\eta^*$ be a non decreasing slowly varying function such that $\lim_{x \to 0} \eta^*(x) = 0$ and such that it is possible to choose a sequence $m$ that satisfies (6). Then, for $t_n$ as in (1),*

$$\liminf_{n \to \infty} \inf_{\widehat{\alpha}_n} \sup_{L \in SV(\eta^*)} \sup_{\alpha \in (-\delta, \delta)} \mathbb{E}_{\alpha,L}[\eta^*(t_n)^{-1} |\widehat{\alpha}_n - \alpha|] = 1 . \tag{8}$$

*Remark 0.2.* This corollary means that the GPH estimator achieves the optimal rate of convergence, up to the exact constant over the class $SV(\eta^*)$ when $\eta^*$ is slowly varying. This implies in particular that, contrarily to the second order regularly varying case, there is no loss of efficiency of the GPH estimator with respect to the GSE. This happens because in the slowly varying case, the bias term dominates the stochastic term if the bandwidth parameter $m$ satisfies (5). This result is not completely devoid of practical importance, since when the rate of convergence of an estimator is logarithmic in the number of observations, constants do matter.

*Example 3.1 (Example 2.2 continued).* If $L(x) = \log^\rho(1/x)\widetilde{L}(x)$, where $\widetilde{L} \in SV(Cx^\beta)$ for some $\rho >, \beta > 0$ and $C > 0$, then $\sum_{k=1}^{m} v_{m,k} \log(L(x_k)) \sim \rho m \log^{-1}(x_m)$. Choosing $m = \log^{1+\delta}(n)$ yields (5), (6) and $\log(n)(\widehat{\alpha}(m) - \alpha)$ converges in probability to $\rho$.

*Proof of Theorem 3.1.* Define $\mathscr{E}_k = \log\{x_k^\alpha I_k / L(x_k)\}$. The deviation of the GPH estimator can be split into a stochastic term and a bias term:

$$\widehat{\alpha}(m) - \alpha = -s_m^{-2} \sum_{k=1}^m v_{m,k} \mathscr{E}_k - s_m^{-2} \sum_{k=1}^m v_{m,k} \log(L(x_k)). \tag{9}$$

Applying Lemma 5.2, we obtain the following bound:

$$\mathbb{E}\left[\left\{\sum_{k=1}^m v_{m,k} \log(\mathscr{E}_k)\right\}^2\right] \leq C(\delta, \eta^*) m. \tag{10}$$

The bias term is dealt with by applying Lemma 5.1 which yields

$$\left|\sum_{k=1}^m v_{m,k} \log(L(x_k))\right| \leq m\eta^*(x_m)\{1 + o(1)\}, \tag{11}$$

uniformly with respect to $|\eta| \leq \eta^*$. Choosing $m$ as in (5) yields (7).                    ◇

# 4 Bandwidth selection

In any semiparametric procedure, the main issue is the bandwidth selection, here the number $m$ of Fourier frequencies used in the regression. Many methods for choosing $m$ have been suggested, all assuming some kind of second order regular variation of the spectral density at 0. In Figures 1- 3 below, the difficulty of choosing $m$ is illustrated, at least visually. In each case, the values of the GPH estimator are plotted against the bandwidth $m$, for values of $m$ between 10 and 500 and sample size 1000.

In Figure 1 the data is a simulated Gaussian ARFIMA(0,$d$,0). The spectral density $f$ of an ARFIMA(0,$d$,0) process is defined by $f(x) = \sigma^2 |1 - e^{ix}|^{-2d}/(2\pi)$, where $\sigma^2$ is the innovation variance. Thus it is second order regularly varying at zero and satisfies $f(x) = x^{-\alpha}(C + O(x^2))$ with $\alpha = 2d$. The optimal choice of the bandwidth is of order $n^{4/5}$ and the semiparametric optimal rate of convergence is $n^{2/5}$. Of course, it is a regular parametric model, so a $\sqrt{n}$ consistent estimator is possible if the model is known, but this is not the present framework. The data in Figure 2 comes from an ARFIMA(0,$d$,0) observed in additive Gaussian white noise with variance $\tau^2$. The spectral density of the observation is then

$$\frac{\sigma^2}{2\pi}|1 - e^{ix}|^{-2d} + \frac{\tau^2}{2\pi} = \frac{\sigma^2}{2\pi}|1 - e^{ix}|^{-2d}\left\{1 + \frac{\tau^2}{\sigma^2}|1 - e^{ix}|^{2d}\right\}.$$

It is thus second order regularly varying at 0 and the optimal rate of convergence is $n^{2d/(4d+1)}$, with optimal bandwidth choice of order $n^{4d/(4d+1)}$. In Figures 1 and 2 the outer lines are the 95% confidence interval based on the central limit theorem for the GPH estimator of $d = \alpha/2$. See [13].

A visual inspection of Figure 1 leaves little doubt that the true value of $d$ is close to .4. In Figure 2, it is harder to see that the correct range for the bandwidth is somewhere between 50 and 100. As it appears here, the estimator is always negatively biased for large $m$, and this may lead to underestimating the value of $d$. Methods to correct this bias (when the correct model is known) have been proposed and investigated by [7] and [9], but again this is not the framework considered here.

Finally, in Figure 3, the GPH estimator is computed for a Gaussian process with autocovariance function $\gamma(k) = 1/(k+1)$ and spectral density $\log|1 - e^{ix}|^2$. The true value of $\alpha$ is zero, but the spectral density is infinite at zero and slowly varying. The plot $\widehat{d}(m)$ is completely misleading. This picture is similar to what is called the Hill "horror plot" in tail index estimation. The confidence bounds are not drawn here because there are meaningless. See Example 3.1.

There has recently been a very important literature on methods to improve the rate of convergence and/or the bias of estimators of the long memory parameter, always under the assumption of second order regular variation. If this assumption fails, all these methods will be incorrect. It is not clear if it is possible to find a realistic method to choose the bandwidth $m$ that would still be valid without second order regular variation. It might be of interest to investigate a test of second order regular variation of the spectral density.

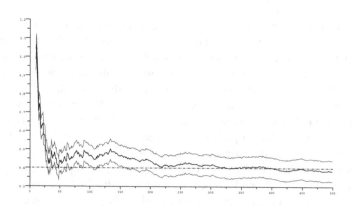

**Fig. 1** GPH estimator for ARFIMA(0,.4,0)

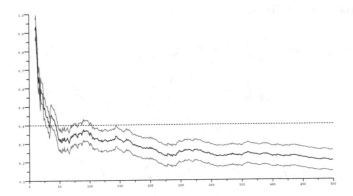

**Fig. 2** GPH estimator for ARFIMA(0,.4,0)+ noise

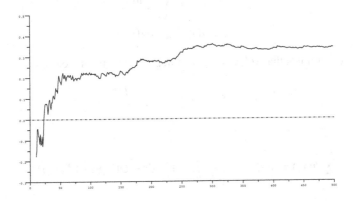

**Fig. 3** GPH "horror" plot

## 5 Technical results

**Lemma 5.1.** *Let $\eta^*$ be a non decreasing slowly varying function on $(0,\pi]$ such that $\lim_{s\to 0}\eta^*(s)=0$. Let $\eta$ be a measurable function on $(0,\pi]$ such that $|\eta|\leq\eta^*$, and define $h(x)=-\int_x^\pi \frac{\eta(s)}{s}\,ds$ and $h^*(x)=\int_x^\pi \frac{\eta^*(s)}{s}\,ds$. Then, for any non decreasing sequence $m\leq n$,*

$$\left|\sum_{k=1}^m v_m(k)h(x_k)\right| \leq m\eta^*(x_m) + O(\log^2(m)\eta^*(x_m) + \log(m)h^*(x_m)), \qquad (12)$$

*uniformly with respect to $|\eta|\leq\eta^*$.*

*Proof.* Since $\eta$ is slowly varying, the function $h$ is also slowly varying and satisfies $\lim_{x \to 0} \eta(x)/h(x) = 0$. Then,

$$\sum_{k=1}^{m} h(x_k) = \frac{n}{2\pi} \int_0^{x_m} h(s) \, ds + \frac{n}{2\pi} \sum_{k=1}^{m} \int_{x_{k-1}}^{x_k} \{h(x_k) - h(s)\} \, ds$$

$$= \frac{n}{2\pi} \int_0^{x_m} h(s) \, ds + \frac{n}{2\pi} \sum_{k=1}^{m} \int_{x_{k-1}}^{x_k} \int_s^{x_k} \frac{\eta(t)}{t} \, dt \, ds .$$

Thus, for $|\eta| \leq \eta^*$ and $\eta^*$ increasing,

$$\left| \sum_{k=1}^{m} h(x_k) - \frac{n}{2\pi} \int_0^{x_m} h(s) \, ds \right| \leq \eta^*(x_m) \frac{n}{2\pi} \sum_{k=1}^{m} \int_{x_{k-1}}^{x_k} \int_s^{x_k} \frac{dt}{t} \, ds$$

$$= \eta^*(x_m) \left\{ \sum_{k=1}^{m} \log(x_k) - \frac{n}{2\pi} \int_0^{x_m} \log(s) \, ds \right\}$$

$$= \eta^*(x_m) \left\{ \sum_{k=1}^{m} \log(k) - m \log(m) + m \right\}$$

$$= O(\eta^*(x_m) \log(m)) .$$

By definition, it holds that $xh'(x) = \eta(x)$. Integration by parts yield

$$\frac{n}{2\pi} \int_0^{x_m} h(s) \, ds = m h(x_m) - \frac{n}{2\pi} \int_0^{x_m} \eta(s) \, ds . \tag{13}$$

Thus,

$$\sum_{k=1}^{m} h(x_k) = m h(x_m) - \frac{n}{2\pi} \int_0^{x_m} \eta(s) \, ds + O(\log(m) \eta^*(x_m)) . \tag{14}$$

Similarly, we have:

$$\sum_{k=1}^{m} h(x_k) \log(x) = m h(x_m) \log(x_m) - \frac{n}{2\pi} \int_0^{x_m} \{\eta(s) \log(s) + h(s)\} \, dx$$

$$+ O(\log(m) \{\log(x_m) \eta^*(x_m) + h^*(x_m)\}) . \tag{15}$$

By definition of $v_m(k)$, we have:

$$v_m(k) = \log(k) - \frac{1}{m} \sum_{j=1}^{m} \log(j) = \log(x_k) - \log(x_m) + 1 + O\left( \frac{\log(m)}{m} \right) .$$

Hence, applying (13), (14) and (15), we obtain

$$\sum_{k=1}^{m} v_m(k)h(x_k)$$

$$= \sum_{k=1}^{m} \log(x_k)h(x_k) - \{\log(x_m) - 1 + O(\log(m)/m)\} \sum_{k=1}^{m} h(x_k)$$

$$= \frac{n}{2\pi} \int_0^{x_m} \eta(s)\{\log(x_m) - \log(s)\}\,ds$$

$$+ O(\log(m)\{\eta^*(x_m)\log(x_m) + h^*(x_m)\}) .$$

Finally, since $|\eta| \le \eta^*$ and $\eta^*$ is non decreasing, we obtain

$$\frac{n}{2\pi} \left| \int_0^{x_m} \eta(s)\{\log(x_m) - \log(s)\}\,ds \right|$$

$$\le \frac{n}{2\pi}\eta^*(x_m) \int_0^{x_m} \int_s^{x_m} \frac{1}{t}\,dt\,ds = m\eta^*(x_m).$$

This yields (12).                                                   ◇

**Lemma 5.2.** *Let $\eta^*$ be a non decreasing slowly varying function such that $\lim_{x \to 0} \eta^*(x) = 0$. Let $X$ be a* Gaussian *process with spectral density $f(x) = x^{-\alpha}L(x)$, where $\alpha \in [-\delta, \delta]$ and $L \in SV(\eta^*)$. Let $\gamma = 0.577216...$ denote Euler's constant. Then, for all $n$ and all $k, j$ such that $0 < x_k, x_j \le \pi/2$,*

$$|\mathbb{E}[\log(\mathcal{E}_k)] - \gamma| + \left| \mathbb{E}[\log^2(\mathcal{E}_k)] - \frac{\pi^2}{6} \right| \le C(\delta, \eta^*)\log(1+k)k^{-1},$$

$$|\mathbb{E}[(\log(\mathcal{E}_k) - \gamma)(\log(\mathcal{E}_j) - \gamma)]| \le C(\delta, \eta^*)\log^2(j)k^{-2}.$$

**Proof of Lemma 5.2**

It is well known (see for instance [8], [11], [14]) that the bounds of Lemma 5.2 are consequences of the covariance inequality for functions of Gaussian vectors of [1, Lemma 1] and of the following bounds. For all $n$ and all $k, j$ such that $0 < |x_k| \le |x_j| \le \pi/2$,

$$|\text{cov}(d_{X,k}, d_{X,j})| + |\text{cov}(d_{X,k}, \bar{d}_{X,j}) - f(x_k)\delta_{k,j}|$$

$$\le C(\delta, \eta^*)\sqrt{f(x_k)f(x_j)}\log(j)k^{-1} .$$

Such bounds have been obtained when the spectral density is second order regularly varying. We prove these bounds under our assumptions that do not imply second order regular variation. Denote $D_n(x) = (2\pi n)^{-1/2} \sum_{t=1}^{n} e^{-itx}$. Then

$$\text{cov}(d_{X,k}, d_{X,j}) = \int_{-\pi}^{\pi} f(x)D_n(x_k - x)D_n(x_j + x)dx.$$

Recall that by definition of the class $SV(\eta^*)$, there exists a function $\eta$ such that $|\eta| < \eta^*$ and $L(x) = L(\pi)\exp\{-\int_x^\pi s^{-1}\eta(s)ds\}$. Since only ratio $L(x)/L(\pi)$ are involved in the bounds, without loss of generality, we can assume that $L(\pi) = 1$. We first prove that for all $k$ such that $x_k \leq \pi/2$,

$$\left| \int_{-\pi}^\pi \left( \frac{f(x)}{f(x_k)} - 1 \right) |D_n(x_k - x)|^2 \, dx \right| \leq C(\delta, \eta^*) \log(k) k^{-1} . \tag{16}$$

Since $L \in SV(\eta^*)$, the functions $x^\varepsilon L(x)$ and $x^\varepsilon L^{-1}(x)$ are bounded for any $\varepsilon > 0$ and

$$\sup_{x \in [0,\pi]} x^\delta (L(x) + L^{-1}(x)) \leq C(\eta^*, \delta) , \tag{17}$$

$$\sup_{\alpha \in [-1+\delta, 1-\delta]} \int_{-\pi}^\pi f(x) \, dx \leq C(\eta^*, \delta) . \tag{18}$$

Since $\eta^*$ is increasing, for all $0 < x < y \leq \pi/2$, it holds that

$$|f(x) - f(y)| = |x^{-\alpha} L(x) - y^{-\alpha} L(y)|$$
$$\leq \int_x^y |\alpha - \eta(s)| s^{-\alpha-1} L(s) \, ds$$
$$\leq \int_x^y (1 + \eta^*(\pi)) s^{-\alpha-1} L(s) \, ds .$$

Since $\alpha \in [-1+\delta, 1-\delta]$, $x^{-\alpha-1} L(x)$ is decreasing. Hence

$$|f(x) - f(y)| \leq C(\eta^*, \delta) x^{-1} f(x)(y - x) . \tag{19}$$

Define $F_n(x) = |D_n(x)|^2$ (the Fejér kernel). We have

$$\sup_{\pi/2 \leq |x| \leq \pi} |F_n(x_k - x)| = O(n^{-1}) , \quad \int_{-\pi}^\pi F_n(x) \, dx = 1 , \tag{20}$$

$$\int_{-\pi}^\pi D_n(y+x) D_n(z-x) \, dx = (2\pi n)^{-1/2} D_n(y+z) . \tag{21}$$

From now on, $C$ will denote a generic constant which depends only on $\eta^*$, $\delta$ and numerical constants, and whose value may change upon each appearance. Applying (17), (18) and (20), we obtain

$$\int_{\pi/2 \leq |x| \leq \pi} |f^{-1}(x_k) f(x) - 1| F_n(x - x_k) \, dx \leq C n^{-1} (f^{-1}(x_k) + 1) \leq C k^{-1} .$$

The integral over $[-\pi/2, \pi/2]$ is split into integrals over $[-\pi/2, -x_k/2] \cup [2x_k, \pi/2]$, $[-x_k/2, x_k/2]$ and $[x_k/2, 2x_k]$. If $x \in [-\pi/2, -x_k/2] \cup [2x_k, \pi/2]$, then $F_n(x - x_k) \leq C n^{-1} x^{-2}$. Hence, applying Karamata's Theorem (cf. [2, Theorem 1.5.8]), we obtain:

$$\int_{-\pi/2}^{-x_k/2} + \int_{-\pi/2}^{-x_k/2} f(x)F_n(x)\,dx$$

$$\leq Cn^{-1}\int_{x_k/2}^{\pi/2} x^{-\alpha-2}L(x)\,dx \leq Cn^{-1}x_k^{-\alpha-1}L(x_k) \leq Ck^{-1}f(x_k)\,,$$

$$\int_{-\pi/2}^{-x_k/2} + \int_{-\pi/2}^{-x_k/2} F_n(x)\,dx \leq Cn^{-1}\int_{x_k/2}^{\infty} x^{-2}\,dx \leq Cn^{-1}x_k^{-1} \leq Ck^{-1}\,.$$

For $x \in [-x_k/2, x_k/2]$, $F_n(x_k - x) \leq n^{-1}x_k^{-2}$. Thus, applying again Karamata's Theorem, we obtain:

$$\int_{-x_k/2}^{x_k/2} f(x)F_n(x-x_k)\,dx \leq Cn^{-1}x_k^{-2}\int_{-x_k/2}^{x_k/2} x^{-\alpha}L(x)\,dx$$

$$\leq Cn^{-1}x_k^{-2}x_k^{-\alpha+1}L(x_k) \leq Ck^{-1}f(x_k)\,,$$

$$\int_{-x_k/2}^{x_k/2} F_n(x-x_k)\,dx \leq Cn^{-1}x_k^{-1} \leq Ck^{-1}\,.$$

Applying (19) and the bound $\int_{-x_k/2}^{x_k} |x|F_n(x)|\,dx \leq Cn^{-1}\log(k)$, we obtain:

$$\int_{x_k/2}^{2x_k} |f(x) - f(x_k)|F_n(x-x_k)\,dx$$

$$\leq Cx_k^{-\alpha-1}L(x_k/2)\int_{x_k/2}^{2x_k} |x - x_k|F_n(x-x_k)|\,dx \leq Cf(x_k)k^{-1}\log(k)\,.$$

This proves (16). We now prove that all $k, j$ such that $0 < x_k \neq |x_j| \leq \pi/2$,

$$\left| \int_{-\pi}^{\pi} \left( \frac{f(x)}{f(x_k)} - 1 \right) D_n(x_k - x)\overline{D_n(x_j - x)}\,dx \right|$$

$$+ \left| \int_{-\pi}^{\pi} \left( \frac{f(x)}{f(x_k)} - 1 \right) D_n(x_k - x)D_n(x_j - x)\,dx \right|$$

$$\leq C(\delta, \eta^*)\log(k \vee |j|)(k \wedge |j|)^{-1}. \quad (22)$$

Define $E_{n,k,j}(x) := D_n(x_k - x)\overline{D_n(x_j - x)}$. Since $0 \leq x_k, x_j \leq \pi/2$, for $\pi/2 \leq |x| \leq \pi$, we have $|E_{n,k,j}(x)| \leq Cn^{-1}$. Hence, as above,

$$\int_{\pi/2 \leq |x| \leq \pi} |f^{-1}(x_k)f(x) - 1|\,dx \leq Cn^{-1}(x_k^{\alpha}L^{-1}(x_k) + 1) \leq Ck^{-1}.$$

We first consider the case $k < j$ and we split the integral over $[-\pi/2, \pi/2]$ into integrals over $[-\pi/2, -x_k/2] \cup [2x_j, \pi/2]$, $[-x_k/2, x_k/2]$, $[x_k/2, (x_k + x_j)/2]$, $[(x_k + x_j)/2, 2x_j]$, denoted respectively $I_1, I_2, I_3$ and $I_4$.
- The bound for the integral over $[-\pi/2, -x_k/2] \cup [2x_j, \pi/2]$ is obtained as above (in the case $k = j$) since $|E_{n,k,j}| \leq Cn^{-1}x^{-2}$. Hence $|I_1| \leq Ck^{-1}$.
- For $x \in [-x_k/2, x_k/2]$, $|E_{n,k,j}(x)| \leq Cn^{-1}x_k^{-2}$, hence we get the same bound: $|I_2| \leq$

$Ck^{-1}$.

• To bound $I_3$, we note that on the interval $[x_k/2, (x_k + x_j)/2]$,

$$|E_{n,k,j}(x)| \leq Cn^{1/2}(j-k)|D_n(x-x_k)|,$$

and $n^{1/2}|x - x_k||D_n(x - x_k|$ is uniformly bounded. Hence, applying (19), we obtain

$$|I_3| \leq C(j-k)^{-1}x_k^{-1}x_j \leq Ck^{-1}.$$

• The bound for $I_4$ is obtained similarly: $|I_4| \leq Ck^{-1}\log(j)$.
• To obtain the bound in the case $x_j < x_k$, the interval $[-\pi, \pi]$ is split into $[-\pi, -\pi/2]$ $\cup [\pi/2\pi]$ $[-\pi/2, -x_k/2] \cup [2x_k, \pi/2]$, $[-x_k/2, x_j/2]$, $[x_j/2, (x_k + x_j)/2]$ and $[(x_k + x_j)/2, 2x_k]$. The arguments are the same except on the interval $[-x_k/2, x_j/2]$ where a slight modification of the argument is necessary. On this interval, it still holds that $|E_{n,k,j}(x)| \leq n^{-1}x_k^{-2}$. Moreover, $x^\delta L(x)$ can be assumed increasing on $[0, x_k/2]$, and we obtain:

$$\int_{-x_k/2}^{x_j/2} x^{-\alpha}L(x)\,dx \leq x_k^\delta L(x_k)\int_{-x_k/2}^{x_j/2} x^{-\alpha-\delta}\,dx \leq Cx_k^{-\alpha+1}L(x_k).$$

The rest of the argument remains unchanged.
• To obtain the bound in the case $x_j < 0 < x_k$, the interval $[-\pi, \pi]$ is split into $[-\pi, -\pi/2] \cup [\pi/2, \pi]$ $[-\pi/2, 2x_j] \cup [2x_k, \pi/2]$, $[2x_j, -x_k/2]$, $[-x_k/2, x_k/2]$ and $[x_k/2, 2x_k]$ and the same arguments are applied. $\diamond$

## References

[1] Arcones, M.A. (1994) Limit theorems for nonlinear functionals of a stationary Gaussian sequence of vectors. *The Annals of Probability*, **22** (4), 2242–2274.
[2] Bingham, N.H., Goldie, C-M, and Teugels, J-L. (1989) *Regular variation*, volume 27 of *Encyclopedia of Mathematics and its Applications*. Cambridge University Press, Cambridge.
[3] Drees, H. (1998) Optimal rates of convergence for estimates of the extreme value index. *The Annals of Statistics*, **26**, 434–448.
[4] Geweke, J. and Porter-Hudak, S. (1983) The estimation and application of long memory time series models. *Journal of Time Series Analysis*, **4**, 221–238.
[5] Giraitis, L., Robinson, P.M. and Samarov, A. (1997) Rate optimal semiparametric estimation of the memory parameter of the Gaussian time series with long range dependence. *Journal of Time Series Analysis*, **18**, 49–61.
[6] Hall, P. and Welsh, A.H. (1984) Best attainable rates of convergence for estimates of parameters of regular variation. *The Annals of Statistics*, **12**, 1079–1084.
[7] Hurvich, C.M. and Ray, B.K. (2003) The local whittle estimator of long-memory stochastic volatility. *Journal of Financial Econometrics*, **1**, 445–470.

[8] Hurvich, C.M., Deo, R. and Brodsky, J. (1998) The mean squared error of Geweke and Porter-Hudak's estimator of the memory parameter of a long-memory time series. *Journal of Time Series Analysis*, **19**, 19–46.

[9] Hurvich, C.M., Moulines, E. and Soulier, P. (2005) Estimating long memory in volatility. *Econometrica*, **73** (4), 1283–1328.

[10] Kuensch. , H.R. (1987) Statistical aspects of self-similar processes. In Yu.A. Prohorov and V.V. Sazonov (eds), *Proceedings of the first World Congres of the Bernoulli Society*, volume 1, pages 67–74. Utrecht, VNU Science Press.

[11] Moulines, E. and Philippe Soulier, P. (1999) Broadband log-periodogram regression of time series with long-range dependence. *The Annals of Statistics*, **27**, 1415–1439.

[12] Robinson. P.M. (1995a) Gaussian semiparametric estimation of long range dependence. *The Annals of Statistics*, **23**, 1630–1661

[13] Robinson. P.M. (1995b) Log-periodogram regression of time series with long range dependence. *The Annals of Statistics*, **23**, 1048–1072.

[14] Soulier, P. (2001) Moment bounds and central limit theorem for functions of Gaussian vectors. *Statistics & Probability Letters*, **54**, 193–203.

# Harmonic analysis tools for statistical inference in the spectral domain

Florin Avram, Nikolai Leonenko, and Ludmila Sakhno

**Abstract** We present here an extension of the theorem on asymptotic behavior of Fejér graph integrals stated in [7] to the case of integrals with more general kernels which allow for tapering. As a corollary, asymptotic normality results for tapered estimators are derived.

## 1 Introduction

In the present paper we aim at extensions of limit theory for Fejér matroid integrals developed in the series of papers [1]–[7] which allow for wider area of statistical applications.

Our object of interest are integrals of the form

$$
\begin{aligned}
J_T &= J_T(M, f_e, e = 1, ..., E) \\
&= \int_{\lambda_1, ..., \lambda_E \in S} f_1(\lambda_1) f_2(\lambda_2) ... f_E(\lambda_E) \prod_{v=1}^{V} K_T(u_v) \prod_{e=1}^{E} \mu(d\lambda_e),
\end{aligned}
\tag{1}
$$

where

Florin Avram
Département de Mathématiques,
Université de Pau et des Pays de l'Adour,
Avenue de l'Université - BP 1155 64013 Pau Cedex France e-mail: Florin.Avram@univ-Pau.fr

Nikolai Leonenko
Cardiff School of Mathematics,
Cardiff University,
Senghennydd Road, Cardiff,CF24 4AG, UK e-mail: LeonenkoN@Cardiff.ac.uk

Ludmila Sakhno
Department of Probability Theory and Mathematical Statistics,
Kyiv National Taras Shevchenko University, Ukraine e-mail: lms@univ.kiev.ua

P. Doukhan et al. (eds.), *Dependence in Probability and Statistics*,
Lecture Notes in Statistics 200, DOI 10.1007/978-3-642-14104-1_4,
© Springer-Verlag Berlin Heidelberg 2010

- $(S, d\mu)$ denotes either $\mathbb{R}^d$, or the torus $\Pi = [-\pi, \pi]^d$ with the Lebesgue measure; $f_e(\lambda) : S \to \mathbb{R}$, $e = 1, \ldots, E$, is a set of functions satisfying integrability conditions

$$f_e \in \mathbf{L}_{p_e}(S, d\mu), \text{ for some } p_e : 1 \le p_e \le \infty, e = 1, \ldots, E; \quad (2)$$

- the variables $\lambda_i$ and $u_v$ are related as

$$(u_1, \ldots, u_V)' = M(\lambda_1, \ldots, \lambda_E)', \quad (3)$$

with $M$ being a matrix of dimensions $V \times E$ (with integer coefficients in the case when $S = [-\pi, \pi]^d$); in the above formula we mean that each $d$-dimensional $u_k$, $k = 1, \ldots, V$, is representable as a linear combination of $d$-dimensional vectors $\lambda_l$, $l = 1, \ldots, E$, with coefficients $M_{k,l}$: $u_k = \sum_{l=1}^{E} M_{k,l} \lambda_l$ (that is, as multiplication of $k$-th row of $M$ by the column $(\lambda_1, \ldots, \lambda_E)'$); such notational convention is for keeping the transparent analogy with the case $d = 1$;

- $K_T(u)$ is a kernel depending on the parameter $T$.

Spaces $\mathbf{L}_p(d\mu)$ appearing above are defined in the following way:

$$\mathbf{L}_p(d\mu) = \begin{cases} L_p(d\mu) & \text{if } 1 \le p < \infty, \\ C_0 & \text{if } p = \infty, \end{cases}$$

where $L_p(d\mu) = \{f : \int_S |f(\lambda)|^p \mu(d\lambda) < \infty\}$, that is, the only difference in comparison with the usual definition of spaces $L_p(d\mu)$, $1 \le p \le \infty$, is that for $p = \infty$ it will be supposed that $\mathbf{L}_\infty$ is the space $C_0$ of all continuous functions which "vanish at infinity", that is, functions $f$ such that for every $\varepsilon > 0$ there exists a compact set $K$ such that $|f(x)| < \varepsilon$ for all $x$ not in $K$ (in the torus case, when $S = \Pi$, this is just the space $C$). With this definition spaces $\mathbf{L}_p$, $1 \le p \le \infty$, will be the closure of the continuous functions of bounded support under the $L_p$ norm, with $L_\infty$ norm being, as usual, uniform supremum norm (see, e.g., [13], Chapter 3).

Previous studies as well as the most recent one undertaken in [7] have been concerned with the case of Dirichlet type kernels

$$K_T(u) = \Delta_T(u) = \int_{t \in I_T} e^{itu} v(dt), \quad (4)$$

where the integration is taken over domains $I_T$ forming the sequence of increasing dilations of a bounded convex domain $I_1$, $I_T = TI_1$, it was assumed that the domain $I_T$ is a cube $[-T/2, T/2]^d$ taken in $\mathbb{R}^d$, or in $\mathbb{Z}^d$ depending on cases $S = \mathbb{R}^d$ or $S = [-\pi, \pi]^d$, $v(\cdot)$ is the Lebesgue measure or counting measure respectively.

The integral (1) with $K_T(u)$ given by (4) is called the Fejér matroid integral (corresponding to the matroid represented by the matrix $M$); for the particular case when $M$ is the incidence matrix of a directed graph, with $V$ vertices and $E$ edges, the construction (1) is called the Fejér graph integral.

We present here an extension of the theorem on asymptotic behavior for Fejér graph integrals, stated in [7], for the case of integrals with more general kernels $K_T(u)$ depending on parameter $T$.

For this we only need the kernels $K_T(u)$ to satisfy the very simple conditions, demanding them (i) to be $\delta$-type kernels and (ii) to obey some boundedness condition with majorant of the form $const \cdot T^a$ with an appropriate exponent $a$ (note that in fact proper bounds in (ii) will lead to (i)). And then all the reasonings from [7] can be applied unaltered to the integrals (1), the main tools in this powerful analytic approach being the Hölder-Young-Brascamp-Lieb inequality and the resolution of certain combinatorial graph optimization problem specific to each particular dependence structure of arguments in (1) given by matrix $M$ via (3).

## 2 Motivation

Problems of statistical estimation of characteristics of random fields (in parametric and nonparametric settings) often require consideration of quadratic forms

$$Q_T = Q_T(g) = \int_{t,s \in I_T} q(t-s)g(X(t),X(s))v(dt)v(ds),$$

based on observations of a measurable, mean square continuous, stationary zero-mean random field $X(t)$, $t \in \mathbb{R}^d$ (or $t \in \mathbb{Z}^d$) over domains $I_T$, with $I_T$ and $v(dt)$ being as in formula (4), and with an appropriate function $g$, for which the above Lebesgue integral exists.

One important particular case is such quadratic forms which can be written as linear integral functionals of the empirical periodogram. Let $\varphi$ be an integrable function, $q(t) = \widehat{\varphi}(t)$ its Fourier transform, and $I_T(\lambda) = (2\pi T)^{-d}\left|\int_{t \in I_T} X(t)e^{it\lambda}v(dt)\right|^2$ be the periodogram of the second order and $g(u,v) = uv$. Then

$$J_T(\varphi) := \int_S \varphi(\lambda)I_T(\lambda)d\lambda = T^{-d}Q_T. \tag{5}$$

One of the classical approaches to derive the central limit theorem for (5) consists in computing the cumulants of (5) and evaluating their asymptotic behavior. In the case of Gaussian or linear fields these cumulants appear to be of the form of the integral (1) with $K_T(u)$ given by (4) and matrix $M$ corresponding to a cyclic graph in the Gaussian case and being of more complicated form for linear fields case (we refer for more detail to [7]).

Consider now the situation where the kernels $K_T(u)$ in (1) are of the form

$$K_T(u) = \int_{I_T} h_T(t)e^{-iut}v(dt), \tag{6}$$

with a certain function $h_T(t)$ described below.

In statistical analysis in spectral domain very often instead of the original data $X(t)$, $t \in I_T$, the tapered data $h_T(t)X(t)$, $t \in I_T$, are used, usually the taper is of the form

$$h_T(t) = \prod_{i=1}^{d} h\left(\frac{t_i}{T}\right), \quad d \geq 1, \tag{7}$$

where $h$ is a function of bounded variation with a bounded support: $h(t) = 0$ for $|t| > 1$ (or $h(t) = 0$ for $t \notin [0,1]$).

Benefits of tapering the data have been widely reported in the literature, e.g., tapers help to reduce leakage effects, especially when the spectral density contains high peaks. Also, the use of tapers leads to the bias reduction, which is especially important when dealing with spatial data: tapers can help to fight the so-called "edge effects" (see, e.g. [10], [12]).

The estimation of the spectral density and spectral functionals are based on the tapered periodogram

$$I_{2,T}(\lambda) = I_{2,T}^{h}(\lambda) = (2\pi H_{2,T}(0))^{-1} d_T^{h}(\lambda) d_T^{h}(-\lambda), \tag{8}$$

where finite Fourier transform

$$d_T^{h}(\lambda) = \sum_{t \in I_T} h_T(t) X(t) e^{-i\lambda t}, \tag{9}$$

and

$$H_{k,T}(\lambda) = \sum_{t \in I_T} h_T^{k}(t) e^{-i\lambda t}, \tag{10}$$

are the so-called spectral windows, and it is supposed that $H_{2,T}(0) \neq 0$. (Let us consider discrete case in the rest of this section, that is, consider a field $X(t)$, $t \in \mathbb{Z}^d$, $I_T \subset \mathbb{Z}^d$, and $\lambda \in [-\pi, \pi]^d$ in the above formulas.)

We have for the cumulants of the finite Fourier transform

$$\text{cum}(d_T(\lambda_1), ..., d_T(\lambda_k)) = \tag{11}$$

$$\int_{\Pi^{k-1}} f_k(\gamma_1, ..., \gamma_{k-1}) H_{1,T}(\lambda_1 - \gamma_1) ... H_{1,T}(\lambda_{k-1} - \gamma_{k-1})$$

$$\times H_{1,T}\left(\lambda_k + \sum_{1}^{k-1} \gamma_j\right) d\gamma_1 ... d\gamma_{k-1},$$

and for the cumulants of the sample spectral functional of the second order

$$J_{2,T}(\varphi) = \int_{\Pi} \varphi(\lambda) I_{2,T}(\lambda) d\lambda, \tag{12}$$

we obtain the following expression, using the product theorem for cumulats (see, e.g., [8] for the case $d = 1$):

$$\text{cum}\,(J_{2,T}(\varphi_1),...,J_{2,T}(\varphi_k)) = (2\pi)^{k-1} H_{2k,T}(0)(H_{2,T}(0))^{-k} \tag{13}$$

$$\times \sum_{v=(v_1,...,v_p)} \int_{\Pi^k} \prod_{j=1}^{k} \varphi_j(\alpha_j) \int_{\Pi^{2k-p}} \prod_{i=1}^{p} f_{|v_i|}(\gamma_{v_i})$$

$$\times \prod_{j=1}^{k} H_{1,T}(\alpha_j - \gamma_{2j-1}) H_{1,T}(-\alpha_j - \gamma_{2j}) \prod_{l=1}^{p} \delta\left(\sum_{j\in v_l} \gamma_j\right) d\gamma\, d\alpha_1...d\alpha_k,$$

where the summation is taken over all indecomposable partitions $v$ of the set of indices from the table T$= \{r_1,...,r_k\}$ with rows $r_i = (2i-1,2i)$, for $v = (i_1,...,i_l)$ we denote $\tilde{v} = (i_1,...,i_{l-1})$. In the above formulas $f_k(\gamma_1,...,\gamma_{k-1})$, $\gamma_i \in \Pi$, are the spectral densities of the field $X(t)$ (which we suppose to exist for all $k \geq 2$) and $\delta(x) = 1$ for $x = 0$ and zero otherwise.

Let us have a look at a particular case of a linear random field $X(t), t \in \mathbb{Z}^d$, with a square integrable kernel $\hat{a}(t)$, that is the filed

$$X(t) = \sum_{u\in\mathbb{Z}^d} \hat{a}(t-u)\xi(u), \quad \sum_{u\in\mathbb{Z}^d} \hat{a}^2(u) < \infty, \, t \in \mathbb{Z}^d, \tag{14}$$

where $\xi(u), u \in \mathbb{Z}^d$, is a independent random variables indexed by $\mathbb{Z}^d$ with $\mathbb{E}\xi(0) = 0$ and such that $E|\xi(0)|^k < \infty$, $k = 1,2,....$ In this case the spectral densities of the field $X(t)$ are of the following form

$$f_k(\lambda_1,...,\lambda_{k-1}) = d_k\, a(-\sum_{i=1}^{k-1}\lambda_i) \prod_{i=1}^{k-1} a(\lambda_i) = d_k \prod_{i=1}^{k} a(\lambda_i)\delta(\sum_{j=1}^{k}\lambda_j), \tag{15}$$

where $d_k$ is the $k$-th cumulant of $\xi(0)$, that is, in particular,

$$d_2 = \mathbb{E}\xi(0)^2, \quad d_4 = \mathbb{E}(\xi(0)^4) - 2[\mathbb{E}(\xi(0)^2)]^2, \tag{16}$$

$\hat{a}(t)$ in (14) are Fourier coefficients for $a(\lambda)$ in (15).

We reveal in the formula (11) the integral of the form (1) with the cyclic product of kernels $K_T(u) = H_{1,T}(u)$ ($M$ corresponds to a cyclic graph), and in the formula (13) – the integral of the form (1) with more complicated dependence structure.

*Note 0.1.* While in the non-tapered case ($h(t) \equiv 1$) the spectral functional of the second order (12) can be written as

$$J_{2,T}(\varphi) = T^{-d}Q_T, \tag{17}$$

with the quadratic form $Q_T = \sum_{t\in I_T}\sum_{s\in I_T} q(t-s)X(t)X(s)$ (where $q(t) = \hat{\varphi}(t)$ is the Fourier transform of $\varphi$), in the tapered case this spectral functional is represented by means of the quadratic form

$$Q_T^h = \sum_{t\in I_T}\sum_{s\in I_T} h(t)h(s)q(t-s)X(t)X(s).$$

In the case of Gaussian field $X(t)$ with the spectral density $f(\lambda)$, analogously to the non-tapered case, the cumulants of $Q_T^h$ can be expressed as

$$c_k = \text{cum}(Q_T^h, ..., Q_T^h) = \text{const } \text{Tr}[(T_T^h(\varphi)T_T^h(f))^k], \qquad (18)$$

where Tr denotes the trace and $T_T^h(\cdot)$ is the "tapered Toeplitz matrix" defined for an integrable function $g$ (with Fourier transform $\widehat{g}$) as

$$T_T^h(g) = (h(t)h(s)\widehat{g}(t-s), \; t,s \in I_T).$$

In the left hand side of (18) we have again a particular case of the integral (1), with cyclic product of kernels (6).

In [11] the following result on the norms of tapered Toeplitz matrix was obtained (in the one-dimensional non-tapered case this result was stated in [1]):

**Proposition 2.1** *Let* $\psi \in L_p(S, d\mu)$, $S = [-\pi, \pi]^d$, $1 \le p \le \infty$. *Then*

$$\left\| T_T^h(\psi) \right\|_p \le T^{d/p} \|\psi\|_p.$$

Here the norms ($p$-Schatten norms) for a linear operator $A$ (with adjoint $A^*$) on a finite dimensional Euclidian space $E$ are defined as follows:

$$\|A\|_p = [\text{Tr}\{(AA^*)^{p/2}\}]^{1/p}, \text{ for } 1 \le p < \infty; \quad \|A\|_\infty = \sup_{\|x\|_E=1} \{(AA^*)^{1/2}x\},$$

where $\|x\|_E$ is the Euclidian norm (the square root of the scalar product of $x$ with itself).

We will state a more general result in our Theorem 3.1 below.

# 3 Main result

Consider the integral (1) with matrix $M$ of dimension $V \times E$ representing a (vectorial) matroid with the rank function $r(A)$, that is, the function, which gives the rank of any set of columns $A$ in $M$. We recall that in particular case of graphic matroids, $M$ is the incidence matrix of a directed graph. To each matroid one may associate a dual matroid with a rank function $r^*(A) = |A| - r(M) + r(M - A)$. A dual matroid is associated with a matrix $M^*$ whose rows span the space orthogonal to the rows of $M$, dimension of $M^*$ is $C \times E$, where $C = E - r(M)$. We denote $co(M) = V - r(M)$ and note that in particular case when $M$ is the incidence matrix of a graph $G$ with $V$ vertices and $E$ edges, $C$ is the maximal number of independent cycles in $G$, $co(M)$ is the number of connected components in $G$. We refer to [7] (and references therein) for any notion unexplained here.

We introduce the following assumptions on kernels $K_T(u)$:

**Assumption 1**

$$\|K_T(u)\|_p \le C_p T^{d(1-\frac{1}{p})}.$$

**Assumption 2** *The kernels*

$$K^*(\gamma_1,...,\gamma_{k-1}) = T^{-d}K_T(\gamma_1)...K_T(\gamma_{k-1})K_T\left(-\sum_1^{k-1}\gamma_j\right),$$

*are approximate identities for convolution, as defined, e.g., in [13]. This implies that for any bounded function $f(x_1,...,x_{k-1})$ continuous at point $(y_1,...,y_{k-1})$ the following holds:*

$$\lim_{T\to\infty}\int_{S^{k-1}} f(x_1-y_1,...,x_{k-1}-y_{k-1})K^*(x_1,...,x_{k-1})\prod_{i=1}^{k-1}\mu(dx_i) = f(y_1,...,y_{k-1}).$$

These assumptions enable us to state the following result, generalizing Theorem 3.1 from [7] (with completely the same reasonings for the proof as in [7]).

Analogously to [7] we will consider the case of graphical matroids and also impose the following assumption on $M$.

**Assumption 3** *For every row $l$ of the matrix $M$, one has $r(M) = r(M_l)$, where $M_l$ is the matrix with the row $l$ removed.*

The main points of the limit result presented below are that:

(1) Upper bounds the integrals (1) involve a specific function $\alpha_M(z)$, related to the graph structure represented by the matrix $M$ and to the integrability indices $p_e$, which is defined as the value of a graph optimization problem. Details of its computation can be found in ([7]).

(2) Under certain conditions on integrability indices $p_e$ (the Hölder-Young-Brascamp-Lieb conditions (21)) necessary to ensure the existence of the limiting integral, the following convergence holds as $T \to \infty$:

$$T^{-d\,co(M)} J_T(M,f_1,...,f_E) \to \int_{S^C} f_1(\lambda_1)f_2(\lambda_2)...f_E(\lambda_E)\prod_{c=1}^{C}\mu(dy_c), \qquad (19)$$

where $(\lambda_1,...\lambda_E) = (y_1,...,y_C)M^*$ (with every $\lambda_e$ reduced modulo $[-\pi,\pi]^d$ in the torus case), $M^*$ being any matrix whose rows span the space orthogonal to the rows of $M$, and $C$ being the rank of $M^*$. Informally, the kernels disappear in the limit, giving rise to the "dual matroid" $M^*$.

(3) When the Hölder-Young-Brascamp-Lieb conditions do not hold, then, see part c) of the theorem below, the normalizing by the factor $T^{d\,\alpha_M(z)}$ appearing in part a) will lead to a zero limit.

**Theorem 3.1.** *Suppose that $f_e \in L_{p_e}(d\mu)$ for part a), and $f_e \in \mathbf{L}_{p_e}(d\mu)$ for parts b) c), and set $z = (p_1^{-1},...,p_E^{-1})$.*

*Let $J_T = J_T(M, f_1, ..., f_E)$ denote the integral of the form (1), the kernels $K_T(u)$ satisfy assumptions 1 and 2, and the matrix $M$ satisfies assumption 3.*
*Then:*

*a)*

$$J_T(M, f_1, ..., f_E) \leq c_M T^{d\,\alpha_M(z)} \prod_{e=1}^{E} \|f_e\|_{p_e}, \qquad (20)$$

*where $c_M$ is some constant independent of $z$ and*

- *in the case $S = [-\pi, \pi]^d$:*

$$\alpha_M(z) = co(M) + \max_{A \subset 1, ..., E} [\sum_{j \in A} z_j - r^*(A)],$$

- *in the case $S = \mathbb{R}^d$:*

$$\alpha_M^c(z) = co(M) + \sum_{j=1}^{co(M)} (\sum_{e \in G_j} z_e - C_j)_+,$$

*where $G_j$, $C_j$ denote the set of columns and the number of cycles belonging to the $j$'th component in $M$.*

*b) If $\alpha_M(z) = V - r(M) = co(M)$ or, equivalently, if*

- *in the case $S = [-\pi, \pi]^d \sum_{j \in A} z_j \leq r^*(A)$, $\forall A$,* $\qquad (21)$

- *in the case $S = \mathbb{R}^d$, in addition $\sum_{e \in G_k} z_e = C_k, k = 1, ..., co(M)$,*

*where $G_k, C_k$ denote respectively the $k$'th component of the graph and its number of cycles, then*

$$\lim_{T \to \infty} \frac{J_T(M, f_1, ..., f_E)}{T^{d\,co(M)}} = \mathscr{J}(M^*, f_1, ..., f_E), \qquad (22)$$

*where*

$$\mathscr{J}(M^*, f_1, ..., f_E) = \int_{S^C} f_1(\lambda_1) f_2(\lambda_2) ... f_E(\lambda_E) \prod_{c=1}^{C} \mu(dy_c), \qquad (23)$$

*where $(\lambda_1, ... \lambda_E) = (y_1, ..., y_C)M^*$ (with every $\lambda_e$ reduced modulo $[-\pi, \pi]^d$ in the case when $S = [-\pi, \pi]^d$), where $C$ denotes the rank of the dual matroid $M^*$.*
*c) If a strict inequality $\alpha_M(z) > co(M)$ holds, then the inequality (20) may be strengthened to:*

$$J_T(M) = o(T^{d\,\alpha_M(z)}).$$

## 4 Applications and discussion

We present here central limit theorems which can be obtained via methods of cumulants as easy consequences of Theorem 3.1.

Consider the sample spectral functionals (12) based on the tapered periodogram.

Useful properties of spectral windows $H_{k,T}(\lambda)$, $\lambda \in \Pi$, $d = 1$, had been stated in [8], in particular, their convolution properties and upper bounds. These results can be extended for $d \geq 1$. We suppose in what follows that in (7) $h : [0,1] \to R$ is a function of bounded variation. The key facts are:

(i) Bounds for $H_{k,T}(\lambda)$ can be given as

$$|H_{k,T}(\lambda)| \leq const \prod_{i=1}^{d} L_T(\lambda_i), \qquad (24)$$

where the functions $L_T(\lambda)$ is $2\pi$-periodic extension of the function

$$L_T^*(\lambda) = \begin{cases} T, & |\alpha| \leq \frac{1}{T}, \\ \frac{1}{|\alpha|}, & \frac{1}{T} < |\alpha| \leq \pi. \end{cases} \qquad (25)$$

(ii) The kernels

$$(2\pi)^{k-1} \left(H_{k,T}(0)\right)^{-1} H_{1,T}(\gamma_1) ... H_{1,T}(\gamma_{k-1}) H_{1,T}\left(-\sum_{1}^{k-1} \gamma_j\right), \gamma_i \in \Pi, \qquad (26)$$

are approximate identities for convolution.

(iii) From Lemma 1 [8] we have $\int_{\Pi} L_T(u)^p du \leq const \, T^{p-1}$, $p > 1$, which leads to the estimates for norms of $H_{k,T}(\lambda)$:

$$\|H_{k,T}(\lambda)\|_p \leq C T^{d(1-\frac{1}{p})}, \; p > 1.$$

Due to the above properties, Theorem 3.1 holds for the integrals (1) with $S = [-\pi, \pi]^d$ and the kernels $K_T(u)$ given by (6), we need only take into account normalizing factors for the kernels (26), which will produce the additional constant multiplier ("tapering factor") in the left hand side of (22) of the form $H_{co(M)} = \{\int_0^1 h^{co(M)}(x)dx\}^d$.

Therefore, for the random field $X(t)$, $t \in \mathbb{Z}^d$, we can analyze the behavior of cumulants of the empirical spectral functional $J_{2,T}(\varphi)$ (12) based on the tapered periodogram, and deduce central limit theorems. We obtain the following result as a simple corollary of Theorem 3.1.

**Theorem 4.1.** *Let $X(t)$, $t \in \mathbb{Z}^d$, be a linear random field given by (14) with a square integrable kernel $\widehat{a}(t)$, and i.i.d. r.v. $\xi(u)$, $u \in \mathbb{Z}^d$, are such that $E|\xi(0)|^k < \infty$, $k = 1, 2, ...$ Assume that the spectral density of the second order $f(\lambda) = (2\pi)^d |a(\lambda)|^2 \in L_p$ and $\varphi(\lambda) \in L_q$, where $p$ and $q$ satisfy:*

$$\frac{1}{p} + \frac{1}{q} \le \frac{1}{2}.$$

Let $J_{2,T}(\varphi)$ is given by (12), and $h(t)$ is of bounded variation, with support on $[0,1]$. Then

(i)

$$T^{-d/2} \{J_{2,T}(\varphi) - EJ_{2,T}(\varphi)\} \to N(0, \sigma^2), \quad T \to \infty,$$

(ii) moreover, if the taper function $h(t)$ is in $C^2([0,1])$ and $f(\lambda)$ or $\varphi(\lambda)$ is in $C^2(\Pi)$ (with $C^2(\cdot)$ being the class of twice continuously differentiable functions) then, for $d = 1,2,3$,

$$T^{-d/2} \{J_{2,T}(\varphi) - J_2(\varphi)\} \to N(0, \sigma^2), \quad T \to \infty,$$

where

$$J_{2,T}(\varphi) = \int_\Pi \varphi(\lambda) f(\lambda) d\lambda,$$

$$\sigma^2 = (2\pi)^d c(h) \left[ 2d_2^2 \int_\Pi \varphi^2(\lambda) f^2(\lambda) d\lambda + (2\pi)^d d_4 \left( \int_\Pi \varphi(\lambda) f(\lambda) d\lambda \right)^2 \right],$$

with the tapering factor being

$$c(h) = \left[ \left( \int_0^1 h^2(t) dt \right)^{-2} \left( \int_0^1 h^4(t) dt \right) \right]^d, \tag{27}$$

and $d_2$ and $d_4$ are defined by (16).

Note that while in the non-tapered case we can state only (i) (see, e.g., Theorem 4.1 in [7]), in the tapered case, due to admissible order of bias $EJ_{2,T}(\varphi) - J_2(\varphi)$ for $d = 1,2,3$ (see, e.g. [10], [12]), we have also CLT (ii), which is important for the problems of statistical estimation, for example, if one needs to derive the asymptotic normality of the so-called quasi-likelihood estimates. That is, in particular, the results formulated in Sections 4.2 and 4.3 in [7] under the assumption $d = 1$ can be extended for the case of random linear fields on $\mathbb{Z}^d$ when $d = 1,2,3$.

In the general (nonlinear, non-Gaussian) case we can obtain the analogous result, under more restrictive conditions, however, it demonstrates again that application of Theorem 3.1 can simplify significantly the reasonings for the proof.

**Theorem 4.2.** Let $X(t)$, $t \in \mathbb{Z}^d$, be a random field, for which all spectral densities exist and are bounded. Let $J_{2,T}(\varphi)$ is given by (12), with $\varphi(\lambda) \in L_2$, and $h(t)$ is of bounded variation, with support on $[0,1]$.
    Then

(i)

$$T^{-d/2} \{J_{2,T}(\varphi) - EJ_{2,T}(\varphi)\} \to N(0, \sigma^2), \quad T \to \infty,$$

*(ii) moreover, if the taper function $h(t)$ is in $C^2([0,1])$ and $f(\lambda)$ or $\varphi(\lambda)$ is in $C^2(\Pi)$, then, for $d = 1, 2, 3$,*

$$T^{-d/2}\{J_{2,T}(\varphi) - J_2(\varphi)\} \to N(0, \sigma^2), \quad T \to \infty,$$

*where $J_{2,T}(\varphi)$ is the same as in Theorem 2,*

$$\sigma^2 = (2\pi)^d c(h) \left[ 2 \int_S \varphi^2(\lambda) f^2(\lambda) d\lambda + (2\pi)^d \int_S \varphi(\lambda) \varphi(\mu) f_4(\lambda, -\lambda, \mu) d\lambda d\mu \right],$$

*and the tapering factor $c(h)$ is given by (27).*

As we have noted above evaluation of the integrals (1) is based essentially on application of the Hölder-Young-Brascamp-Lieb inequality which, under prescribed conditions on the integrability indices for a set of functions $f_i \in \mathbf{L}_{p_i}(S, d\mu)$, $i = 1, \ldots, n$, allows to write upper bounds for the integrals of the form

$$\int_{S^m} \prod_{i=1}^k f_i(l_i(x_1, \ldots, x_m)) \prod_{j=1}^m \mu(dx_j),$$

with $l_i : S^m \to S$ being linear functionals.

However more powerful tool is provided by the non-homogeneous Hölder-Young-Brascamp-Lieb inequality, which covers the case when the above functions $f_i$ are defined over the spaces of different dimensions $f_i : S^{n_i} \to R$, say. With such inequality we would have the possibility to evaluate the integrals of the form (1), where the product of kernels $K_T(u)$ of the form (6) is accompanied by a function involving a product of spectral densities of different orders. These integrals appear in calculation of the cumulants of the empirical spectral functional $J_{2,T}(\varphi)$ (see (13)), and also the empirical spectral functionals of higher orders, for the case of a general (nonlinear, non-Gaussian) random field. Some results in this direction were obtained in [8], [9] in the case when $d = 1$. Namely, in [8] the convergence to zero of normalized cumulants (13) was stated under the conditions of boundedness of $\varphi_i$ and integrability conditions on the spectral densities of the form $f_k \in L_4(\Pi^{k-1})$, $k \geq 2$. In [9] more refined conditions of integrability for the spectral densities were formulated like $f_k \in L_{g(k)}(\Pi^{k-1})$ with some $g(k)$ given in explicit form. The proofs are based on the Hölder inequality and bounds (24) and convolution properties of the function (25). We believe that the mentioned results are particular cases of more general theory – asymptotic theory for Fejér matroid integrals of more general form than (1), which can be developed basing on the non-homogeneous Hölder-Young-Brascamp-Lieb inequality. We will address this topic in further investigations.

**Acknowledgements** The authors are grateful to the referee for detailed comments and suggestions to improve the presentation. The work is partly supported by the EPSRC grant RCMT 119 and by the Welsh Institute of Mathematics and Computational Sciences, and the Commission of the European Communities grant PIRSES-GA-2008-230804.

# References

[1] Avram, F., On Bilinear Forms in Gaussian Random Variables and Toeplitz Matrices. *Probab. Theory Related Fields* **79** (1988) 37–45.

[2] Avram, F., Brown, L. A Generalized Hölder Inequality and a Generalized Szegö Theorem. *Proceedings of the American Math. Soc.* **107** (1989) 687–695.

[3] Avram, F., Taqqu, M.S. Hölder's Inequality for Functions of Linearly Dependent Arguments. *SIAM J. Math. Anal.* **20** (1989) 1484–1489.

[4] Avram, F. Generalized Szegö Theorems and asymptotics of cumulants by graphical methods. *Transactions of the American Math. Soc.* **330** (1992) 637–649.

[5] Avram, F., Fox, R. Central limit theorems for sums of Wick products of stationary sequences. *Transactions of the American Math. Soc.* **330** (1992) 651–663.

[6] Avram, F., Taqqu, M.S. On a Szegö type limit theorem and the asymptotic theory of random sums, integrals and quadratic forms. *Dependence in probability and statistics*, Lecture Notes in Statist., 187, Springer, New York, (2006), 259–286.

[7] Avram, F., Leonenko, N., Sakhno, L. On a Szegö type limit theorem, the Hölder-Young-Brascamp-Lieb inequality, and the asymptotic theory of integrals and quadratic forms of stationary fields. *ESAIM: Probab. Statist.* (2009), to appear.

[8] Dahlhaus, R. Spectral analysis with tapered data. *J. Time Series Anal.* **4** (1983) 163–175.

[9] Dahlhaus, R. A functional limit theorem for tapered empirical spectral functions. *Stochastic Process. Appl.* **19** (1985) 135–149.

[10] Dahlhaus, R., Künsch, H., Edge effects and efficient parameter estimation for stationary random fields, *Biometrika*, 74 (1987), 877-882. 39–81.

[11] Doukhan, P., Leon, J.R., Soulier, P. Central and non central limit theorems for quadratic forms of a strongly dependent Gaussian filed. *REBRAPE* **10** (1996) 205-223.

[12] Guyon, X., *Random Fields on a Network: Modelling, Statistics and Applications.* Springer, New York, 1995.

[13] Rudin, W. *Real and Complex Analysis.* McGraw-Hill, London, New York (1970).

# On the impact of the number of vanishing moments on the dependence structures of compound Poisson motion and fractional Brownian motion in multifractal time

Béatrice Vedel, Herwig Wendt, Patrice Abry, and Stéphane Jaffard

**Abstract** From a theoretical perspective, scale invariance, or simply scaling, can fruitfully be modeled with classes of multifractal stochastic processes, designed from positive multiplicative martingales (or cascades). From a practical perspective, scaling in real-world data is often analyzed by means of multiresolution quantities. The present contribution focuses on three different types of such multiresolution quantities, namely increment, wavelet and Leader coefficients, as well as on a specific multifractal processes, referred to as Infinitely Divisible Motions and fractional Brownian motion in multifractal time. It aims at studying, both analytically and by numerical simulations, the impact of varying the number of vanishing moments of the mother wavelet and the order of the increments on the decay rate of the (higher order) covariance functions of the ($q$-th power of the absolute values of these) multiresolution coefficients. The key result obtained here consist of the fact that, though it fastens the decay of the covariance functions, as is the case for fractional Brownian motions, increasing the number of vanishing moments of the mother wavelet or the order of the increments does not induce any faster decay for the (higher order) covariance functions.

Béatrice Vedel
Université de Bretagne Sud, Université Européenne de Bretagne, Campus de Tohannic, BP 573, 56017 Vannes, e-mail: beatrice.vedel@univ-ubs.fr

Herwig Wendt, Patrice Abry
ENS Lyon, CNRS UMR 5672, 46 allée d'Italie, 69364 Lyon cedex, e-mail: herwig.wendt@ens-lyon.fr, patrice.abry@ens-lyon.fr

Stéphane Jaffard
Université Paris Est, CNRS, 61, avenue du Général de Gaulle, 94010 Créteil, e-mail: jaffard@univ-paris12.fr
.

P. Doukhan et al. (eds.), *Dependence in Probability and Statistics*,
Lecture Notes in Statistics 200, DOI 10.1007/978-3-642-14104-1_5,
© Springer-Verlag Berlin Heidelberg 2010

# 1 Motivation

**Scale invariance.** Scale invariance, or simply scaling, is a paradigm nowadays commonly used to model and describe empirical data produced by a large variety of different applications [31]. Scale invariance consists of the idea that the data being analyzed do not possess any characteristic scale (of time or space). Often, instead of being formulated directly on the data $X(t)$, scale invariance is expressed through *multiresolution coefficients*, $T_X(a,t)$, such as increments or wavelet coefficients, computed from the data, and depending jointly on the position $t$ and on the analysis scale $a$. Practically, scaling is defined as power law behaviors of the ($q$-th order) moments of the (absolute values of the) $T_X(a,t)$ with respect to the analysis scale $a$:

$$\frac{1}{n_a}\sum_{k=1}^{n_a}|T_X(a,ak)|^q \simeq c_q a^{\zeta(q)}, \tag{1}$$

for a given range of scales $a \in [a_m, a_M]$, $a_M/a_m \gg 1$, and for some (statistical) orders $q$.

Performing scaling analyses on data mostly amounts to testing the adequacy of Eq. (1) and to estimate the corresponding scaling exponents $\zeta(q)$.

**Fractional Brownian motion and vanishing moments.** Fractional Brownian motion (fBm) has been among the first stochastic process used to model scaling properties in empirical data and still serves as the central reference. It consists of the only Gaussian self similar process with stationary increments, see [24] and references therein. It is mostly controlled by its self-similarity parameter, $0 < H < 1$, and its scaling exponents behave as a linear function of $q$: $\zeta(q) = qH$.

For $1/2 < H < 1$, the increment process of fBm is characterized by a so-called *long range dependence* structure, or *long memory* [24, 6]. This property significantly impairs the accurate estimation of the process parameters.

Two seminal contributions [12, 13] (see also [27]) showed that the wavelet coefficients of fBm are characterized by a *short range* correlation structure, as soon as the number $N_\psi$ of vanishing moments (cf. Section 3, for definition) of the analyzing mother wavelet $\psi_0$ is such that:

$$N_\psi \geq 2H + 1. \tag{2}$$

This *decorrelation* or *whitening* property of the wavelet coefficients implies, because of the Gaussian nature of fBm, that the entire dependence structure is turned *short range*. This significantly simplifies the estimation of the model parameters, and notably that of the self-similarity parameter $H$ [2, 28, 32, 10]. It has, later, been shown that this decorrelation property is also effective when using higher order increments (increments of increments...) [16], which equivalently amounts to increasing the number of vanishing moments of the mother wavelet (cf. Section 3). This key role of $N_\psi$, together with the possibility that it can be easily tuned by practitioners, has constituted the fundamental motivation for the systematic and popular use of wavelet transforms for the analysis of empirical data which are likely to possess scaling properties.

**Multiplicative martingales and vanishing moments.** Often in applications, scaling exponents are found to depart from the linear behavior $qH$ associated to fBm. To account for this, following Mandelbrot's seminal works (e.g., [20]), multiplicative cascades (or more technically multiplicative martingales) have received considerable interests in modeling scaling in applications. Notably, they are considered as reference processes to model multifractal properties in data, a particular instance of scaling. Often, it has been considered heuristically by practitioners that the benefits of conducting wavelet analyses over such models were equivalent to those observed when analyzing fBm. Despite its being of crucial importance, this issue received little attention at a theoretical level (see a contrario [3, 15]). This can partly be explained by the fact that Mandelbrot's multiplicative cascades, that remain up to the years 2000, the most (if not the only) practically used processes to model multifractal scaling properties in data, possess involved statistical characteristics. Hence, the derivation of the dependence structure of their wavelet coefficients has been considered difficult to obtain analytically. In turn, this prevents a theoretical analysis of the statistical properties of the scaling estimation procedure based on Eq. (1) (see a contrario [23] for one of the very few theoretical result). More recently, a new type of multiplicative martingales, referred to as compound Poisson cascades and motions, have been introduced [5]. They were later generalized to infinitely divisible cascades and motions and fractional Brownian motions in multifractal time [21, 25, 22, 8, 9]. Such processes are regarded as fruitful alternatives to the original and celebrated Mandelbrot's cascades, as they enable to define processes with both known multifractal properties and stationary increments. Moreover, they are very easy to simulate. Therefore, they provide practitioners with relevant and efficient tools both for the analysis and the modeling of multifractal properties in real world data. This is why they are studied in details in the present contribution.

Yet, to our knowledge, neither the dependence structures of their wavelet or increment coefficients nor the impact on such structures of varying the number of vanishing moments of the analyzing wavelet, or of the order of the increments, have so far been studied.

**Multiresolution analysis.** To study scale invariance and scaling properties, increments have been the most commonly used multiresolution coefficients and their statistical properties are traditionally and classically analyzed in the literature. However, it is now well known that wavelet coefficients provide relevant, accurate and robust tools for the practical analysis of scale invariance (self-similarity and long range dependence notably, cf. [1]). Furthermore, it has recently been shown [17, 18] that the accurate analysis of multifractal properties requires the use of wavelet Leaders instead of wavelet coefficients or increments.

**Goals and contributions.** The central aim of the present contribution consists of studying the impact of varying the number of vanishing moments of the analyzing wavelet or the order of the increments on the correlation and dependence (or higher order correlation) structures of three different multiresolution coefficients (increment, wavelet and Leader coefficients) for two different classes of multifractal stochastic processes, infinitely divisible motions and fractional Brownian motions in multifractal time. This is achieved by combining analytical studies and numerical analysis. The major contribution is to show that, while increasing the number of vanishing moments of the analyzing wavelet or the order of the increments significantly fasten the decay of the correlation functions of the increment and wavelet coefficients, in a manner comparable to what is observed for fBm, it does not impact at all the decay of higher order correlation functions.

**Outline.** The key definitions and properties of infinitely divisible cascades, motions and fractional Brownian motions in multifractal time are recalled in Section 2. The three different types of multiresolution quantities (increment, wavelet and leader coefficients) are defined and related one to the other in Section 3. This section also briefly recalls scaling exponent estimation procedures. Analytical results regarding the correlation and higher order correlation functions of increment and wavelet coefficients are carefully detailed in Section 4. Results and analyses obtained from numerical simulations for the higher order correlation functions of increment, wavelet and Leader coefficients are reported in Section 5. Conclusions on the impact of varying the number of vanishing moments of the analyzing wavelet or the order of the increment on correlation and on higher order correlation function decays are discussed in Section 6. The proofs of all analytical results are postponed to Section 7.

## 2 Infinitely divisible processes

### 2.1 Infinitely divisible cascade

**Infinitely divisible measure.** Let $G$ be an infinitely divisible distribution [11], with moment generating function

$$\tilde{G}(q) = \exp[-\rho(q)] = \int \exp[qx]dG(x).$$

Let $M$ denote an infinitely divisible, independently scattered random measure distributed by the infinitely divisible distribution $G$, supported by the time-scale half plane $\mathscr{P}^+ = \mathbb{R} \times \mathbb{R}^+$ and associated to its control measure $dm(t,r)$.

Infinitely divisibility with generating function $G$ and control measure $m$ implies that, for any Borel set $\mathscr{E} \in \mathscr{P}^+$, the moment generating function of the random variable $M(\mathscr{E})$ reads, using the Lévy-Khintchine formula:

$$\mathbb{E}[\exp[qM(\mathscr{E})]] = \exp[-\rho(q)m(\mathscr{E})] = \exp\left[-\rho(q)\int_{\mathscr{E}} dm(t,r)\right]. \qquad (3)$$

**Infinitely divisible cascade.**    Let $\mathscr{C}_r(t)$, $0 < r < 1$, denote a so-called truncated influence cone, defined as

$$\mathscr{C}_r(t) = \{(t',r') : r \le r' \le 1, t - r'/2 \le t' < t + r'/2\}.$$

Following the definition of compound Poisson cascades [5] (see paragraph below), it has been proposed in [25, 22, 8, 9] to define infinitely divisible cascades (or noises) as follows.

**Definition 0.1.** An Infinitely Divisible Cascade (or Noise) (IDC) is a family of processes $Q_r(t)$, parameterized by $0 < r < 1$, of the form

$$Q_r(t) = \frac{\exp[M(\mathscr{C}_r(t))]}{\mathbb{E}[\exp[M(\mathscr{C}_r(t))]]} = \exp[\rho(1)m(\mathscr{C}_r(t))]\exp[M(\mathscr{C}_r(t))]. \qquad (4)$$

Let us moreover define

$$\widetilde{\varphi}(q) = \rho(q) - q\rho(1) = -\log\left(\frac{\mathbb{E}[e^{qX}]}{\mathbb{E}[e^X]^q}\right) = -\log\left(\frac{\mathbb{E}[Z^q]}{(\mathbb{E}[Z])^q}\right),$$

whenever defined, with $Z = \exp[X]$ and $X$ distributed according to the infinitely divisible distribution $G$. Obviously, $\widetilde{\varphi}(q)$ is a concave function, with $\widetilde{\varphi}(0) = \widetilde{\varphi}(1) = 0$. Note that the cone of influence is truncated at the scale $r$, so that details of smaller scale are absent in the construction of the cascade. The mathematical difficulty lies in understanding the limit when $r \to 0$, in which case all (fine) scales are present.

**Control measure.**    In the remainder of the text, following [4, 8], the control measure is chosen such that $dm(t,r) = \mu(dr)dt$. The choice of the shift-invariant Lebesgue measure $dt$ will have the consequence that the processes constructed below have stationary increments. Following [4], $\mu(dr)$ is set to $\mu(dr) = c(dr/r^2 + \delta_{\{1\}}(dr))$, where $\delta_{\{1\}}(dr)$ denotes a point mass at $r = 1$.

**Central property.**    With this choice of measure, an infinitely divisible cascade possesses the following key property:

**Proposition 0.1.** [5, 4, 8] Let $Q_r$ be an infinitely divisible cascade, $0 < r < 1$. Then, $\forall q > 0$, such that $\mathbb{E}[Q_r^q(t)] < \infty$, $\forall t \in \mathbb{R}$,

$$\mathbb{E}[Q_r^q(t)] = \exp[-\widetilde{\varphi}(q)m(\mathscr{C}_r(t))] = e^{-\varphi(q)}r^{\varphi(q)}, \qquad (5)$$

where, for ease of notation, $\varphi(q) \equiv c\widetilde{\varphi}(q)$.

**Integral scale.**    By construction, infinitely divisible cascades $Q_r$ are intrinsically tied to a characteristic scale, referred to as the *integral scale*, in the hydrodynamic turbulence literature (cf. e.g., [20]). Essentially, it results from the fact that the influence cone $\mathscr{C}_r(t)$ is limited above by an upper limit $r \le r' \le L$. Traditionally, and without loss of generality, the upper limit is assigned to the reference value $L \equiv 1$, as only the ratio $L/r$ controls the properties of $Q_r$ (cf. [8]). Therefore, in what follows, we study the properties of the different processes defined from $Q_r$ only for $0 \leq t \leq L \equiv 1$.

**Compound Poisson cascade.**    Compound Poisson cascades (CPC) [5] consist of a particular instance within the IDC family. They are obtained by taking for infinitely divisible random measure $M$ a sum of weighted Dirac masses, as follows: One considers a Poisson point process $(t_i, r_i)$ with control measure $dm(t,r)$, and one associates to each of the points $(t_i, r_i)$ positive weights which are i.i.d. random variables $W_i$. Then, the definition Eq. (4) reads [5]:

$$Q_r(t) = \frac{\exp[\sum_{(t_i,r_i)\in\mathscr{C}_r(t)} \log W_i]}{\mathbb{E}[\exp[\sum_{(t_i,r_i)\in\mathscr{C}_r(t)} \log W_i]]} = \frac{\prod_{(t_i,r_i)\in\mathscr{C}_r(t)} W_i}{\mathbb{E}[\prod_{(t_i,r_i)\in\mathscr{C}_r(t)} W_i]}. \tag{6}$$

When the control measure $dm(t,r)$ is chosen as above, CPC satisfies Eq. (5) with [5]:

$$\varphi(q) = (1 - \mathbb{E}[W^q]) - q(1 - \mathbb{E}W), \tag{7}$$

whenever defined.

## 2.2 Infinitely divisible motion

The following remarkable result states that the random cascades $Q_r(t)$ have a weak limit when $r \to 0$ (i.e. their distribution functions have a pointwise limit).

**Proposition 0.2.** *Let $Q_r(t)$ denotes an infinitely divisible cascade and $A_r(t)$ be its mass distribution process, with $A_r(t) = \int_0^t Q_r(u)du$. There exists a càdlàg (right-continuous with left limits) process $A(.)$ such that almost surely*

$$A(t) = \lim_{r \to 0} A_r(t),$$

*for all rational $t$ simultaneously. This process $A$ is a well defined process on condition that $\varphi'(1^-) \geq -1$.*

The increasing process $A$ is often referred to as an Infinitely Divisible Motion.

**Remark 1:** Since $Q_r > 0$, all processes $A_r$ and $A$ are non-decreasing and therefore have right and left limits; thus, $A$ can be extended to a càdlàg process defined on $\mathbb{R}$.

**Proposition 0.3.** *[5, 4, 8] The process $A$ is characterized by the following properties:*

*1. It possesses a scaling property of the form*

$$\forall q \in (0, q_c^+), \ \forall t \in (0,1), \qquad \mathbb{E}A(t)^q = C(q)t^{q+\varphi(q)}, \tag{8}$$

*with $C(q) > 0$ and*

$$q_c^+ = \sup \{q \geq 1, q + \varphi(q) - 1 \geq 0\} ; \qquad (9)$$

2. *Its increments are stationary w.r.t. time t, hence, the expectation $\mathbb{E}A(t)A(s)$ necessarily reads $\mathbb{E}A(t)A(s) = f(|t|) + f(|s|) - f(|t-s|)$;*
3. *By combining the two previous items, we immediately obtain the detailed form of $\mathbb{E}A(t)A(s)$:*

$$\mathbb{E}A(t)A(s) = \sigma_A^2(|t|^{2+\varphi(2)} + |s|^{2+\varphi(2)} - |t-s|^{2+\varphi(2)}), \quad |t-s| \leq 1. \qquad (10)$$

*Moreover, the constant $\sigma_A^2$ can be computed explicitly, see [29]:*

$$\sigma_A^2 \equiv C(2) \equiv \mathbb{E}A(1)^2 = ((1 + \varphi(2))(2 + \varphi(2)))^{-1}.$$

4. *The multifractal[1] properties of the sample paths $A(t)$ are entirely controlled by the only function $q + \varphi(q)$. This can be inferred from the results in [5].*

Note moreover that it follows from the above results that:

$$\mathbb{E}A(t)A(s) - \mathbb{E}A(t)\mathbb{E}A(s) = \sigma_A^2(|t|^{2+\varphi(2)} + |s|^{2+\varphi(2)} - |t-s|^{2+\varphi(2)}) - C(1)^2|t||s|,$$

$$\mathbb{E}(A(t+\tau) - A(t))(A(s+\tau) - A(s)) =$$
$$\sigma_A^2(|t-s+\tau|^{2+\varphi(2)} + |t-s-\tau|^{2+\varphi(2)} - 2|t-s|^{2+\varphi(2)}).$$

Because this latter must be a positive definite functions of $t, s$, $A(t)$ is a well defined second order process with stationary increments if and only if $0 < 2 + \varphi(2) \leq 2$ (cf., e.g., [26]). Importantly, this restricts the class of functions $\varphi$ that can enter the definition of the second order process $A$ to those satisfying:

$$-2 < \varphi(2) \leq 0. \qquad (11)$$

From now on, it is further assumed that there exists $q > 1$ such that $q + \varphi(q) - 1 > 0$. This is notably the case for the class of strictly concave functions $\varphi$, of interest here. Moreover, the case $\varphi = 0$ on an interval is excluded (it would correspond to the non interesting constant IDC noise $Q_r(t) = 1$ case). Under such assumptions, the following lemma can be proven (cf. Section 7.2):

**Lemma 2.1.** *Assume $q_c^+ > 2$. For $0 < q < q_c^+$ and for any $t > 0$:*

$$\lim_{r \to 0} \mathbb{E}|A(t) - A_r(t)|^q = 0.$$

---

[1] For a thorough introduction to multifractal analysis, the reader is referred to e.g., [17].

## 2.3 Fractional Brownian motion in multifractal time

The stationary increment infinitely divisible motion $A(t)$ suffers from a severe limitation as far as data modeling is concerned: It is a monotonously increasing process. Following the original idea proposed in [21], it has been proposed to overcome this drawback by considering fractional Brownian motion subordinated to $A(t)$:

**Definition 0.2.** Let $B_H$ denote fBm with self-similarity parameter $0 < H < 1$ and $A(t)$ a stationary increment infinitely divisible motion $A(t)$. The process

$$\mathscr{B}(t) = B_H(A(t)) \tag{12}$$

is referred to as fractional Brownian motion in multifractal time (MF-fBm, in short).

The properties of $\mathscr{B}(t)$ stem from the combination of those of $B_H$ and of the random time change $A$ that was performed:

**Proposition 0.4.**
1. The process $\mathscr{B}(t)$ possesses the following scaling property:

$$\forall 0 < q < q_c^+/H, \quad \mathbb{E}|\mathscr{B}(t)|^q = c_q |t|^{qH + \varphi(qH)}; \tag{13}$$

2. Its increments are stationary with respect to time $t$ ;
3. Combining stationary increments and scaling yields the following covariance function:

$$\mathbb{E}\mathscr{B}(t)\mathscr{B}(s) = \sigma_{\mathscr{B}}^2 \left( |t|^{2H + \varphi(2H)} + |s|^{2H + \varphi(2H)} - |t - s|^{2H + \varphi(2H)} \right), \tag{14}$$

where $\sigma_{\mathscr{B}}^2 = \mathbb{E}|B_H(1)|^2/2$.
4. The multifractal properties of the sample paths $\mathscr{B}(t)$ are entirely controlled by the only function $qH + \varphi(qH)$.

Moreover, Eq. (14) involves a positive semi definite function, therefore $\mathscr{B}$ is is a well defined second order process with stationary increments if and only if [26]: $\forall H \in (0,1)$, $0 < 2H + \varphi(2H) \le 2$. Importantly, this restricts the class of functions $\varphi$ that can enter the definition of $\mathscr{B}$ to those satisfying:

$$\forall q \in (0,2), \ -q < \varphi(q) \le 2 - q. \tag{15}$$

# 3 Multiresolution quantities and scaling parameter estimation

## 3.1 Multiresolution quantities

**Increments.**    The increments of order $P$, of a function $X$, taken at lag $\tau$ are defined as:

$$X_{(\tau,P)}(t) \equiv \sum_{k=0}^{P} (-1)^k \binom{P}{k} X(t - k\tau). \tag{16}$$

**Wavelet coefficients.** The discrete wavelet transform (DWT) coefficients of $X$ are defined as

$$d_X(j,k) = \int_{\mathbb{R}} X(t) \, 2^{-j} \psi_0(2^{-j}t - k) \, dt. \tag{17}$$

The mother-wavelet $\psi_0(t)$ consists of an oscillating reference pattern, chosen such that the collection $\{2^{-j/2}\psi_0(2^{-j}t - k), j \in \mathbb{N}, k \in \mathbb{N}\}$ forms an orthonormal basis of $L^2(\mathbb{R})$. Also, it is characterized by its *number of vanishing moments*: an integer $N_\psi \geq 1$ such that $\forall k = 0, 1, \ldots, N_\psi - 1$, $\int_{\mathbb{R}} t^k \psi_0(t) dt \equiv 0$ and $\int_{\mathbb{R}} t^{N_\psi} \psi_0(t) dt \neq 0$. This $N_\psi$ controls the behavior of the Fourier transform $\widetilde{\Psi}_0$ of $\psi_0$ at origin:

$$|\widetilde{\Psi}_0(v)| \sim |v|^{N_\psi}, \ |v| \to 0. \tag{18}$$

**Wavelet leaders.** In the remainder of the text, we further assume that $\psi_0(t)$ has a compact time support. Let us define dyadic intervals as $\lambda = \lambda_{j,k} = [k2^j, (k+1)2^j)$, and let $3\lambda$ denote the union of the interval $\lambda$ with its 2 adjacent dyadic intervals: $3\lambda_{j,k} = \lambda_{j,k-1} \cup \lambda_{j,k} \cup \lambda_{j,k+1}$. Following [17, 31], wavelet Leaders are defined as (for ease of notations, $L_X(j,k) = L_\lambda$ and $d_X(j,k) = d_\lambda$:

$$L_X(j,k) = L_\lambda = \sup_{\lambda' \subset 3\lambda} |d_{\lambda'}|. \tag{19}$$

The wavelet Leader $L_X(j,k)$ practically consists of the largest wavelet coefficient $|d_X(j',k')|$ at all finer scales $2^{j'} \leq 2^j$ in a narrow time neighborhood. It has been shown theoretically that the analysis of the multifractal properties of sample paths of stochastic processes and particularly the estimation of their scaling parameters is relevant and accurate only if wavelet Leaders, rather than wavelet coefficients or increments, are chosen as multiresolution quantities (cf. [17, 31]).

**Vanishing moments.** Increments, as defined in Eq. (16) above, can be read as wavelet coefficients, obtained from the specific mother-wavelet

$$\psi_0^I(t) = \sum_{k=0}^{P} (-1)^k \binom{P}{k} \delta(t - k)$$

(where $\delta$ is the Dirac mass function):

$$X_{(2^j,P)}(2^j k) \equiv d_X(j,k; \psi_0^I).$$

It is straightforward to check that the number of vanishing moments of this particular mother-wavelet corresponds to the increment order: $N_\psi \equiv P$. This explains why increments are sometimes referred to as wavelet coefficients, obtained from a low regularity mother-wavelet (historically referred to as the *poor man's wavelet* [14]). Moreover, increments are often heuristically regarded as *derivatives of order P*, the same interpretation holds for wavelet coefficients, computed from a mother wavelet with

$$N_\psi = P \geq 1 \tag{20}$$

vanishing moments. Hence, $N_\psi$ and $P$ play similar roles.

**Multiresolution analysis.**   From the definitions above, it is obvious that varying the lag $\tau$ corresponds equivalently to changing the analysis scale $a = 2^j$. Therefore, increments consist of multiresolution quantities, in the same spirit as wavelet coefficients and Leaders do. In the present contribution, we therefore analyze in a common framework the three types of multiresolution quantities, $T_X(a,t)$, increment, wavelet and Leader coefficients: $T_X(2^j, 2^j k)$ will denote either $X_{2^j,P}(2^j k)$, $d_X(j,k)$, or $L_X(j,k)$.

### 3.2 Scaling parameter estimation procedures

Inspired from Eq. (1), classical scaling parameter $\zeta(q)$ estimation procedures are based on linear regressions, over dyadic scales $a_j = 2^{j_1}, \ldots, 2^{j_2}$ ($\Sigma$ stands for $\Sigma_{j=j_1}^{j_2}$, the weights $w_j$ satisfy $\Sigma w_j = 0$ and $\Sigma j w_j = 1$) (cf. [19, 30] for details):

$$\hat{\zeta}(q) = \sum w_j \log_2 \left( \frac{1}{n_j} \sum_{k=1}^{n_j} |T_X(2^j, 2^j k)|^q \right). \tag{21}$$

The analysis of the statistical performance of such estimation procedures requires the knowledge of the multivariate dependence structures of the variables $|T_X(a,t)|^q$. Some aspects of these dependence structure are studied in the next sections.

## 4 Dependence structures of the multiresolution coefficients: analytical study

The aim of the present section is to study analytically, both for infinitely divisible motion and for fractional Brownian motion in multifractal time, $X = A, \mathcal{B}$, the impact of varying $P$ or $N_\psi$ on some aspects of the dependence structures of the $T_X(a,t)$, when the $T_X(a,t)$ are either increments or discrete wavelet coefficients.

Section 4.1 starts with the theoretical analyses of the covariance functions $\mathbb{E}T_X(a,t)T_X(a,s)$.

Section 4.2 continues with the theoretical studies of the *higher order* covariance functions $\mathbb{E}|T_X(a,t)|^q|T_X(a,s)|^q$, for some $q$s: The key point being that the absolute values, $|\cdot|$, consisting of non linear transforms of the $T_X(a,t)$ capture well all aspects of the dependence structure.

All results reported in Sections 4 and 5 are stated for $0 < t, s < 1$, hence for $|t - s| < 1$ (i.e., within the integral scale).

## 4.1 Correlation structures for increment and wavelet coefficients

### 4.1.1 Increments

From Eq. (10) (Eq. (14), resp.), it can be shown that the covariance structure of the increments of order $P$ of Infinitely divisible motion $A$ (Fractional Brownian motion in multifractal time $\mathscr{B}$, resp.) reads:

$$\mathbb{E}A_{(\tau,P)}(t)A_{(\tau,P)}(s) = \sigma_A^2 \sum_{n=0}^{2P} (-1)^n \binom{2P}{n} (t - s + (n - P)\tau)^{2+\varphi(2)}, \qquad (22)$$

$$\mathbb{E}\mathscr{B}_{(\tau,P)}(t)\mathscr{B}_{(\tau,P)}(s) = \sigma_\mathscr{B}^2 \sum_{n=0}^{2P} (-1)^n \binom{2P}{n} (t - s + (n - P)\tau)^{2H+\varphi(2H)}. \quad (23)$$

Taking the limit $\tau \to 0$, i.e., $1 > |t - s| \gg \tau$ yields:

$$\lim_{\tau \to 0} \frac{\mathbb{E}A_{(\tau,P)}(t)A_{(\tau,P)}(s)}{|\tau|^{2P}} = C_P(\varphi(2))|t - s|^{2+\varphi(2)-2P}, \qquad (24)$$

$$\lim_{\tau \to 0} \frac{\mathbb{E}\mathscr{B}_{(\tau,P)}(t)\mathscr{B}_{(\tau,P)}(s)}{|\tau|^{2P}} = C_{P,H}(\varphi(2H))|t - s|^{2H+\varphi(2H)-2P}. \qquad (25)$$

The direct proofs are detailed in [29]. Such results can also be obtained by following the proof of Proposition 0.5 below.

### 4.1.2 Wavelet coefficients

Let $\psi_0$ denote a compact support mother wavelet with $N_\psi$ vanishing moments.

**Proposition 0.5.** Let $j, k, k' \in \mathbb{Z}$, $|k - k'| \leq C2^{-j}$ such that the supports of $\psi_{j,k}$ and $\psi_{j,k'}$ are included in $[0,1]$. Then

$$\lim_{j \to -\infty} \frac{\mathbb{E}d_A(j,k)d_A(j,k')}{2^{2jN_\psi}} \sim O(|2^j k - 2^j k'|)^{(2+\varphi(2)-2N_\psi)}, \qquad (26)$$

$$\lim_{j \to -\infty} \frac{\mathbb{E}d_\mathscr{B}(j,k)d_\mathscr{B}(j,k')}{2^{2jN_\psi}} \sim O(|2^j k - 2^j k'|)^{(2H+\varphi(2H)-2N_\psi)}. \qquad (27)$$

Proof. We follow step by step the calculations conducted in [13] on fractional Brownian motion (note that we use a $L^1$-normalization for the wavelet coefficients instead of the usual $L^2$-normalization, as in [13]).

From the form of the covariance function (cf. Eq. (10)), one obtains:

$$\mathbb{E}d_A(j,k)d_A(j,k') = -\sigma_A^2(2^j)^{2+\varphi(2)}\int\left(R_\psi(1,\tau-(k-k'))|\tau|^{1+\varphi(2)/2}\right)d\tau$$

with

$$R_\psi(\alpha,\tau) = \sqrt{\alpha}\int_{-\infty}^{+\infty}\psi_0(t)\psi_0(\alpha t - \tau)dt$$

the reproducing kernel of $\psi_0$. Rewriting the relation above in the Fourier domain and using Eq. (18) yields:

$$\mathbb{E}d_A(j,k)d_A(j,k') = \sigma_{A,\psi_0}^2(2^j)^{2+\varphi(2)+1}\int_{-\infty}^{+\infty}e^{i\omega(k-k')}\frac{|\Psi_0(\omega)|^2}{|\omega|^{2+\varphi(2)+1}}\frac{d\omega}{2\pi},$$

with $\sigma_{A,\psi_0}^2 = \sigma_A^2 2\sin(\frac{\pi}{2}(2+\varphi(2)))\Gamma(4+2\varphi(2)+1)$, and hence:

$$\mathbb{E}d_A(j,k)d_A(j,k') = 2^{j(2+\varphi(2)}O(|k-k'|^{2+\varphi(2)-2N_\psi}).$$

Now, since $0 \le |k-k'| \le C2^{-j}$ (which means that the support of $\psi_{j,k}$ is centered on a point $t = c2^j k$), we can write $k = 2^{-j}p$, $k' = 2^{-j}p'$ for $|p-p'| \le C$, and

$$\mathbb{E}d_A(j,k)d_A(j,k') = \mathbb{E}d_A(j,2^{-j}p)d_A(j,2^{-j}p') = 2^{2jN_\psi}O(|p-p'|^{2+\varphi(2)-2N_\psi}).$$

The proof for Eq. (27) is similar.

### 4.1.3 Vanishing moments and correlation

These computations show two striking results regarding the impact of varying $N_\psi$ (resp., $P$) on the asymptotic expansion in $2^j$ (resp., $|\tau|$) of the covariance function of the wavelet coefficients (resp., the increments).

-   First, when $2^j \to 0$ (resp., $|\tau| \to 0$), the order of magnitude of the leading term in the asymptotic expansion decreases with $N_\psi$ (resp., $P$) as $2^{2jN_\psi}$ (resp, $|\tau|^{2P}$) when $N_\psi$ (resp., $P$) increases.
-   Second, the leading term in $2^j|k-k'|$ (resp., $|t-s|$) decreases faster when $N_\psi$ (resp., $P$) increases.

To conduct comparisons against results obtained for fractional Brownian motions, this shows first that the limit $|t-s| \to +\infty$ needs to be replaced with the limit $|\tau| \to 0$ and the range $\tau \le |t-s| \le 1$. Then, one observes that the impact of varying $N_\psi$ (resp., $P$) on the correlation structures of the wavelet coefficients (resp., the increments) of Infinitely Divisible Motion and fractional Brownian motion in multifractal time is equivalent, mutatis mutandis, to that obtained for fractional Brownian motion: The larger $N_\psi$ (resp., $P$), the faster the decay of the correlation functions. Such results call for the following comments:

-   This comes as no surprise as both processes $A$ and $\mathscr{B}$ share with $B_H$ two key properties: scale invariance and stationary increments.

- Because $A$ and $\mathscr{B}$ are non Gaussian processes, the derivation of their correlation structures does not induce the knowledge of their dependence (or higher order correlation) structures. This is why functions $\mathbb{E}|T_X(a,t)|^q|T_X(a,s)|^q$ are further analyzed in the next sections.
- No analytical results are available for the correlation of the wavelet Leaders, $\mathbb{E}L_X(j,k)L_X(j,k')$. This is because, while increment and wavelet coefficients are obtained from linear transforms of $X$, the Leaders $L_X(j,k)$ are derived from non linear transforms (as do the $|T_X(a,t)|^q$ in general).
- Because $\varphi(1) \equiv 0$, the key quantity $\varphi(2)$ can be rewritten $\varphi(2) - 2\varphi(1)$, for comparisons with the form of the scaling exponents entering the Theorems proven below.

## 4.2 Higher order correlations for increments

### 4.2.1 First order increments

**Infinitely divisible motion $A$.**     The covariance function for the integer $q$-th power of the first order increments of $A$ can be obtained analytically:

**Theorem 4.1.** *Let $q$ be an integer such that $1 \leq q < q_c^+/2$. There exists $c(q) > 0$ such that, for $0 < t, s < 1$,*

$$\lim_{\tau \to 0} \frac{\mathbb{E}A_{(\tau,1)}{}^q(t)A_{(\tau,1)}{}^q(s)}{c(q)|\tau|^{2(q+\varphi(q))}} = |t-s|^{\varphi(2q)-2\varphi(q)}.$$

*The constant $c(q)$ can be calculated precisely from the formula*

$$c(q)|\tau|^{2(q+\varphi(q))} = (\mathbb{E}A_{(\tau,1)}{}^q(1))^2.$$

The proof of Theorem 4.1 is postponed to Section 7.2.

When $q \notin \mathbb{N}$, an exact result for scaling is not available, yet the following inequalities can be obtained, which show that the exact power behavior obtained for integer $q$ extend to real $q$, at least in the limit $|t-s| \to 0$.

**Theorem 4.2.** *Let $q$ be a real number such that $1 \leq q < q_c^+/2$. There exists $C_1 > 0$ and $C_2 > 0$ depending only on $q$ such that,*

$$C_1|t-s|^{\varphi(2q)-2\varphi(q)} \leq \lim_{\tau \to 0} \frac{\mathbb{E}A_{(\tau,1)}{}^q(t)A_{(\tau,1)}{}^q(s)}{|\tau|^{2(q+\varphi(q))}} \leq C_2|t-s|^{\varphi(2q)-2\varphi(q)}.$$

*In particular,*

$$\varphi(2q) - 2\varphi(q) = \inf\left\{\alpha \in \mathbb{R}; \ \lim_{|t-s|\to 0}\frac{1}{|t-s|^{\alpha}}\lim_{\tau\to 0}\frac{\mathbb{E}A_{(\tau,1)}{}^{q}(t)A_{(\tau,1)}{}^{q}(s)}{|\tau|^{2(q+\varphi(q))}} = +\infty\right\}$$

$$= \sup\left\{\alpha \in \mathbb{R}; \ \lim_{|t-s|\to 0}\frac{1}{|t-s|^{\alpha}}\lim_{\tau\to 0}\frac{\mathbb{E}A_{(\tau,1)}{}^{q}(t)A_{(\tau,1)}{}^{q}(s)}{|\tau|^{2(q+\varphi(q))}} = 0\right\}.$$

This is proven in Section 7.2.

Moreover, the dependence of increments taken at two different analyzing scales, $\tau_1$ and $\tau_2$ is given by the following result, which follows from the proof of Theorem 4.1.

**Corollary 4.1.** *Let* $1 \le q < q_c^+/2$ *be an integer. There exists* $c(q) > 0$ *such that*

$$\lim_{\tau_1,\tau_2\to 0}\frac{\mathbb{E}A_{(\tau_1,1)}^{q}(t)A_{(\tau_2,1)}^{q}(s)}{c(q)|\tau_1\tau_2|^{q+\varphi(q)}} = |t-s|^{\varphi(2q)-2\varphi(q)}.$$

This shows that the power law behavior in Theorem 4.1 for increments taken at the same scale $\tau$ can be extended to increments defined at any two different scales.

**Fractional Brownian motion in multifractal time $\mathscr{B}$**

**Proposition 0.6.** *Let* $1 > H > 1/2$. *There exists two constants* $C_1 > 0$ *and* $C_2 > 0$ *such that,*

$$C_1|t-s|^{\varphi(4H)-2\varphi(2H)} \le \lim_{\tau\to 0}\frac{\mathbb{E}\mathscr{B}_{(\tau,1)}^{2}(t)\mathscr{B}_{(\tau,1)}^{2}(s)}{\tau^{4H+2\varphi(2H)}} \le C_2|t-s|^{\varphi(4H)-2\varphi(2H)}.$$

The proof, which is a consequence of Theorem 4.2, is postponed to Section 7.4.

### 4.2.2 Second order increments

Let us now study the correlation of the squared ($q = 2$) second order ($P = 2$) increments of Infinitely divisible motion $A$ and fractional Brownian motion in multifractal time $\mathscr{B}$.

**Proposition 0.7.** *Let* $\varphi$ *be chosen such that* $q_c > 4$ *(cf. Eq. (9)). There exists* $c > 0$ *such that,*

$$\lim_{\tau\to 0}\frac{\mathbb{E}A_{(\tau,2)}^{2}(t)A_{(\tau,2)}^{2}(s)}{\mathbb{E}A_{(\tau,2)}^{2}(t)\mathbb{E}A_{(\tau,2)}^{2}(s)} = \lim_{\tau\to 0}\frac{\mathbb{E}A_{(\tau,2)}^{2}(t)A_{(\tau,2)}^{2}(s)}{c|\tau|^{2(\varphi(2)+2)}} = |t-s|^{\varphi(4)-2\varphi(2)}.$$

The proof is detailed in Section 7.5.

**Proposition 0.8.** *Let* $\varphi$ *be chosen such that* $q_c > 4$ *(cf. Eq. (9)) and let* $1 > H > 1/2$. *There exists two constants* $C_1 > 0$ *and* $C_2 > 0$ *such that, for all* $0 \le t < s \le 1$,

$$C_1|t-s|^{\varphi(4H)-2\varphi(2H)} \le \lim_{\tau \to 0} \frac{\mathbb{E}\mathscr{B}^2_{(\tau,2)}(t)\mathscr{B}^2_{(\tau,2)}(s)}{\tau^{4H+2\varphi(2H)}} \le C_2|t-s|^{\varphi(4H)-2\varphi(2H)}.$$

The proof is detailed in Section 7.6. This result suggests that the higher order co-variance functions of $\mathscr{B}$ have the same behaviors as those of $A$, replacing $q$ with $qH$.

## 4.3 Role of the order of the increments

Comparing, the results of Theorem 4.1 (for $\mathbb{E}A^2_{\tau,1}(t)A^2_{\tau,1}(s) \equiv \mathbb{E}|A_{\tau,1}(t)|^2|A_{\tau,1}(s)|^2$) versus those of Proposition 0.7 (for $\mathbb{E}A^2_{\tau,2}(t)A^2_{\tau,2}(s) \equiv \mathbb{E}|A_{\tau,2}(t)|^2|A_{\tau,2}(s)|^2$), on one hand, and results of Proposition 0.6 (for $\mathbb{E}\mathscr{B}^2_{\tau,1}(t)\mathscr{B}^2_{\tau,1}(s) \equiv \mathbb{E}|\mathscr{B}_{\tau,1}(t)|^2|B_{\tau,1}(s)|^2$) versus those of Proposition 0.8 (for $\mathbb{E}B^2_{\tau,2}(t)B^2_{\tau,2}(s) \equiv \mathbb{E}|B_{\tau,2}(t)|^2|B_{\tau,2}(s)|^2$), on other hand, yields the first major conclusion: Increasing $P$ from 1 to 2 induces, for the higher order correlation functions, neither a change in the order in $|\tau|$ of the leading term of the asymptotic expansion in $|\tau| \to 0$, nor any faster decay of the coefficient in $|t-s|$ for this leading term. This is in clear contrast with the impact of $P$ on the correlation functions $\mathbb{E}A_{\tau,P}(t)A_{\tau,P}(s)$ and $\mathbb{E}\mathscr{B}_{\tau,P}(t)\mathscr{B}_{\tau,P}(s)$, with $P = 1,2$ (cf. results of Section 4.1).

# 5 Dependence structures of the multiresolution coefficients: Conjectures and numerical studies

## 5.1 Conjectures

The analytical results obtained in Section 4 for the $q$-th ($q = 1,2$) power of the absolute values of the first and second order increments of both processes $X = A, \mathscr{B}$ lead us to formulate the two following conjectures, for the three different multiresolution quantities $T_X(a,t)$ considered here (increment, wavelet and Leader coefficients).

*Conjecture 0.1.* Let $1 \le q < q_c^+/2$. There exists $C_A(q) > 0$ depending only on $q$ such that, for $0 < s - t < 1$, one has

$$\lim_{a \to 0} \frac{\mathbb{E}|T_A(a,t)|^q|T_A(a,s)|^q}{|a|^{2(q+\varphi(q))}} = C_A(q)|t-s|^{\varphi(2q)-2\varphi(q)}.$$

*Conjecture 0.2.* Let $1 \le qH < q_c^+/2$. There exists $C_{\mathscr{B}}(q) > 0$ depending only on $q$ such that, for $0 < s - t < 1$, one has

$$\lim_{a \to 0} \frac{\mathbb{E}|T_{\mathscr{B}}(a,t)|^q |T_{\mathscr{B}}(a,s)|^q}{|a|^{2(qH+\varphi(qH))}} = C_{\mathscr{B}}(q)|t-s|^{\varphi(2qH)-2\varphi(qH)}.$$

The central features of these two conjectures consists of the following facts:

1. the higher order covariance functions decay algebraically (by concavity of the function $\varphi(q)$, for all $q > 0$, the quantity $\varphi(2q) - 2\varphi(q)$ is strictly negative);
2. the scaling exponent of the leading term characterizing the algebraic decay in $|t - s|$ is not modified when the order $P$ of the increments or the number of vanishing moment $N_\psi$ of the mother wavelet are increased.

## 5.2 Numerical simulations

To give substance to these conjectures, formulated after the analytical results obtained in Section 4, the following sets of numerical simulations are performed and analyzed.

### 5.2.1 Simulation set up

Numerical simulations are conducted on compound Poisson motions (i.e., processes $A$ obtained from compound Poisson cascades) rather than on infinitely divisible motions, as the former are much easier to handle from a practical synthesis perspective. The practical synthesis of realizations of the latter is linked with heavy computational and memory costs, which impose severe practical limitations (in maximally possible sample size, for instance). Therefore, they remain barely used in applications. In contrast, realizations of processes $A$ and $\mathscr{B}$ based on compound Poisson cascades are reasonably easy to simulate numerically (cf. [8] for a review). For ease of notations, the corresponding processes $A$ and $\mathscr{B}$ are referred to as CPM and CPM-MF-fBm.

More specifically, we illustrate results with log-Normal multipliers $W = \exp[V]$, where $V \overset{i.i.d.}{\sim} \mathcal{N}(\mu, \sigma^2)$ are Gaussian random variables. In this case, the function $\varphi(q)$ Eq. (7) is given by:

$$\varphi(q) = c\left[\left(1 - \exp\left(\mu q + \frac{\sigma^2}{2}q^2\right)\right) - q\left(1 - \exp\left(\mu + \frac{\sigma^2}{2}\right)\right)\right]. \quad (28)$$

This choice is motivated by the fact that it is often considered in applications [7].

For CPM, numerical results reported here are computed with parameters $c = 1$, $\mu = -0.3$ and $\sigma^2 = 0.04$ (hence $q_c^+/2 \approx 8.92$); $nbreal = 100$ realizations of the process are numerically synthetised, with sample size $N = 2^{21}$. For CPM-MF-fBm, the same parameters are used, with in addition $H = 0.7$ (hence $(q_c^+/H)2 \simeq 12.74$) and the sample size is reduced to $N = 2^{18}$ (as the synthesis of realizations of CPM-MF-fBm is slightly more involved than for CPM, hence limiting the obtainable sample size to a smaller one, as compared to CPM).

From these, the increment, wavelet and Leader coefficients of CPM A and CPM-MF-fBm $\mathscr{B}$ are calculated for a chosen analysis scale $a$. Results are illustrated here for $a = 2^3$, but similar conclusions are drawn for any different $a$. The number of vanishing moments (and increment orders) are set to $P \equiv N_\psi = \{1, 2, 5, 8\}$. The statistical orders $q$ are in the range $0 < q < q_c/2$.

We developed both the synthesis and analysis codes in MATLAB and they are available upon request.

### 5.2.2 Goal and analysis

Following the formulation of Conjectures 0.1 and 0.2 above, the goal of the simulations is to validate the power law decay of the correlation functions of the $q$-th power of the absolute values of the multiresolution quantities and to estimate the corresponding power law exponents $\alpha_A(q, N_\psi) = \varphi(2q) - 2\varphi(q)$ and $\alpha_{\mathscr{B}}(q, N_\psi) = \varphi(2qH) - 2\varphi(qH)$, controlling respectively such decays.

To do so, we make use of the wavelet based spectral estimation procedure documented and validated in [28], whose key features are briefly recalled here. Let $Y$ denote a second order stationary process with covariance function, $\mathbb{E}Y(t)Y(s) \simeq |t - s|^{-\alpha}$. Then, it has been shown that $1/n_j \sum_k^{n_j} d_Y^2(j,k) \simeq 2^{j\alpha}$. Therefore, the parameter $\alpha$ can be estimated from a linear regression in the diagram $\log_2 1/n_j \sum_k^{n_j} d_X^2(j,k)$ vs. $\log_2 2^j = j$:

$$\widehat{\alpha} = \sum_{j=j_1}^{j_2} w_j \log_2 \left( 1/n_j \sum_k^{n_j} d_Y^2(j,k) \right),$$

the weights $w_j$ satisfy $\sum_{j=j_1}^{j_2} w_j = 0$ and $\sum_{j=j_1}^{j_2} j w_j = 1$. This wavelet based estimation procedure is applied to time series consisting of the $q$-th power of the absolute value of increment, wavelet and Leader coefficients of $A$ and $\mathscr{B}$ computed at an arbitrary scale $a$: $Y(t) \equiv |T_X(a,t)|^q$. Estimation is performed with a compact support Daubechies wavelet with 4 vanishing moments.

### 5.2.3 Results and analyses

For increments, wavelet coefficients and Leaders, for all $0 < q < q_c^+$, for all $N_\psi \equiv P$, for both CPM A and CPM-MF-fBm $\mathscr{B}$, satisfactory power laws behaviors can be observed in diagrams

$$\log_2 \left( \frac{1}{n_j} \sum_k^{n_j} d_X^2(j,k) \right) \quad \text{vs.} \quad \log_2 2^j = j.$$

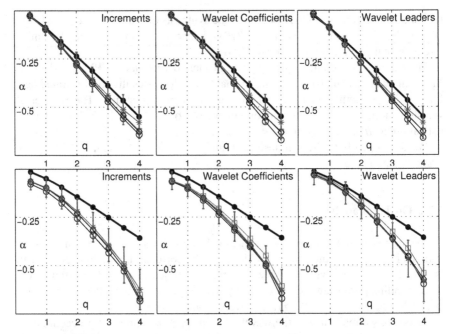

**Fig. 1 Wavelet based estimations of power law exponents for the higher correlation functions.**
Wavelet based estimates (averaged over realizations) of the power law exponents $\widehat{\alpha}_A(q,N_\psi)$ (for
CPM A, top row) and $\widehat{\alpha}_{\mathscr{B}}(q,N_\psi)$ (CPM-MF-fBm $\mathscr{B}$, bottom row) of the higher order correlations
of increments (left), wavelet coefficients (center) and wavelet Leaders (right), as functions of $q$.
The conjectured exponents, $\alpha_A(q,N_\psi) = \varphi(2q) - 2\varphi(q)$ and $\alpha_{\mathscr{B}}(q,N_\psi) = \varphi(2qH) - 2\varphi(qH)$, are
drawn in fat solid black with dots. The symbols { '*' , '◇' , '□' , '○' } correspond to $P \equiv N_\psi = \{1,2,5,8\}$, respectively. The error bars (in red) for $N_\psi = P = 1$ correspond to 95% asymptotic
confidence intervals.

and power scaling exponent can be estimated. These plots are not shown here.

Fig. 1 represents (in bold black solid with dots) the conjectured power law
exponents $\alpha_A(q,N_\psi) = \varphi(2q) - 2\varphi(q)$ (for CPM A, top row) and $\alpha_{\mathscr{B}}(q,N_\psi) = \varphi(2qH) - 2\varphi(qH)$ (CPM-MF-fBm $\mathscr{B}$, bottom row), as a function of $q$, and com-
pares them to averages over realizations of estimates of power law exponents
$\widehat{\alpha}_A(q,N_\psi)$ and $\widehat{\alpha}_{\mathscr{B}}(q,N_\psi)$, for increments (left), wavelet coefficients (center) and
wavelet Leaders (right) with $P \equiv N_\psi = \{1,2,5,8\}$ vanishing moments. Also, for
$N_\psi = 1$, asymptotic 95% error bars are shown. Error bars for other values of $N_\psi$ are
of similar size and omitted for better readability.

The results in Fig. 1 clearly indicate that, both for A and $\mathscr{B}$:

- No significant difference can be observed between the power law exponents
estimated from increments, wavelet coefficients, and wavelet Leaders: The differences between the different estimated scaling exponents are very small, and within
confidence intervals.

- These power law exponents, characterizing the decay of the higher order correlation functions of the multiresolution quantities, practically do not vary when the number of vanishing moments (or the order of the increments) is increased.

- For A, for fixed $q$, notably for small $q$, the differences between the estimated and conjectured scaling exponents remain small, and within confidence intervals. The slight discrepancies for large $q$ values can be interpreted by the fact that only leading order terms are taken into account by Conjecture 0.1, whereas in practice, for finite size realizations, higher order terms may also contribute (cf. Remark 3 in Section 7.5). For $\mathscr{B}$, estimated scaling exponents appear farther from (and smaller than) the conjectured ones. This discrepancy can be explained by the limited sample size for realizations of CPM-MF-fBm (roughly one order of magnitude smaller than those obtainable for A). Therefore, higher order terms are likely to impact more significantly than for A on the asymptotic expansion of $\mathbb{E}|T_{\mathscr{B}}(a,t)|^q|T_{\mathscr{B}}(a,s)|^q$, and hence to impair the accurate estimation of the power law exponent of the leading term.

- Conjectures actually also apply to the range $q \in [0,1]$.

These results and analyses strongly support Conjectures 0.1 and 0.2. Similar results and equivalent conclusions are drawn from other choices of $\varphi(q)$ and from numerical simulations performed directly on infinitely divisible motions (though of lower sample size, and not reported here).

# 6 Discussions and conclusions on the role of the number of vanishing moments:

The theoretical studies (reported in Section 4), together with the numerical analyses (conducted in Section 5) yield the following major conclusions regarding the impact of varying $N_\psi$ or $P$ on the higher order correlations functions of $X_{(\tau,P)}(t)$, $d_X(j,k)$ and $L_X(j,k)$.

- The correlation functions of the increments and wavelet coefficients of A and $\mathscr{B}$ decay faster when $N_\psi \equiv P$ is increased. This effect is equivalent to that obtained for fractional Brownian motion, and results from the facts that $B_H$, A and $\mathscr{B}$ posses a scale invariance property (as in Eqs. (8) or (13)) and stationary increments and wavelet coefficients(and also from the fact that increments and wavelet coefficients consist of linear transform of the processes).

- For the $q$-th power ($0 < q < q_c^+/2$) of Leaders or of the absolute values of the increments or wavelet coefficients, which consist of non linear transforms of processes A and $\mathscr{B}$, $N_\psi$ and $P$ no longer impact the decay of the correlation functions.

- The power law exponents characterizing the algebraic decay of these correlation functions are found to be identical for increment, wavelet and Leader coefficients.

- These power law exponents are conjectured to be controlled by the only function $\varphi(q)$ underlying the construction of the infinitely divisible cascade for A, and, in addition, parameter $H$ for $\mathscr{B}$: $\alpha_A(q,N_\psi) = \varphi(2q) - 2\varphi(q)$ and $\alpha_{\mathscr{B}}(q,N_\psi) =$

$\varphi(2qH) - 2\varphi(qH)$.

- Furthermore, both estimated and predicted power law exponents are larger than $-1$. This is significantly so for small $q$ values. This would be the case for most, if not any, choices of function $\varphi(q)$ commonly used in applications. This reveals very slow (algebraic) decay of the covariance functions, a important characteristic with respect to parameter estimation issues.

After the seminal works on self similar fractional Brownian motions and wavelet coefficients [12, 13] or increments [16], we believe that the present theoretical analyses and numerical results are closing a gap: They shed new lights and significantly renew the understanding of the role of the order of the increments and of the number of vanishing moments of the mother wavelet, for wavelet coefficients and Leaders, with respect to the analysis of scale invariance as modeled by infinitely divisible multifractal processes.

# 7 Proofs

## 7.1 A key lemma

The proofs of the results obtained in the present contribution rely on the key Lemma in [4], restated here. Let $\varphi(\cdot) = \psi(-i\cdot)$ and let $\omega_r$ be defined by $Q_r = e^{\omega_r}$. For $t, t' \geq 0$, one defines:

$$\mathscr{C}_r(t, t') = \mathscr{C}_r(t) \cap \mathscr{C}_r(t')$$

**Lemma 7.1.** *Let $q \in \mathbb{N}^*$, and the vectors $\vec{t_q} = (t_1, t_2, ..., t_q)$ with $t_1 \leq t_2 \leq ... \leq t_q$ and $\vec{p_q} = (p_1, p_2, ..., p_q)$. The characteristic function of the vector $\{w_r(t_m)\}_{1 \leq m \leq q}$ reads:*

$$\mathbb{E}\left(e^{\sum_{m=1}^{q} ip_m M(\mathscr{C}_r(t_m))}\right) = e^{\sum_{j=1}^{q} \sum_{k=1}^{j} \alpha(j,k)\rho_r(t_k - t_j)}$$

*where $M$ is the infinitely divisible, independently scattered random measure used in the construction of $Q_r$,*

$$\rho_r(t) = m(\mathscr{C}_r(0, t)),$$

$$\alpha(j, k) = \psi(r_{k,j}) + \psi(r_{k+1,j-1}) - \psi(r_{k,j-1}) - \psi(r_{k+1,j})$$

*and*

$$r_{k,j} = \begin{cases} \sum_{m=k}^{j} p_m, & \text{for } k \leq j, \\ 0 & \text{for } k > j. \end{cases}$$

*Moreover,*

$$\sum_{j=1}^{q} \sum_{k=1}^{j} \alpha(j, k) = \psi\left(\sum_{k=1}^{q} p_k\right).$$

This can be rewritten as

$$\mathbb{E}Q_r^{p_1}(t_1)Q_r^{p_2}(t_2)...Q_m^{p_m}(t_m) = e^{\sum_{j=1}^{q}\sum_{k=1}^{j}\beta(j,k)\rho_r(t_k-t_j)}$$

with

$$\beta(j,k) = \varphi(r_{k,j}) + \varphi(r_{k+1,j-1}) - \varphi(r_{k,j-1}) - \varphi(r_{k+1,j})$$

and

$$\sum_{j=1}^{q}\sum_{k=1}^{j}\beta(j,k) = \varphi\left(\sum_{k=1}^{q}p_k\right).$$

This form betrays the way the cones $\mathscr{C}_r$ located at $(t_1, t_2, ..., t_q)$ overlap.

## 7.2 Proof of Theorem 4.1

Let us first prove Lemma 2.1. Let $f(q) = q + \varphi(q) - 1$. Since $\varphi$ is concave, so is $f$. Moreover, we assumed that there exists $q > 1$ such that $f(q) > 0$. Since $f$ might only be constant in the neighborhood of its maximum, it implies that $q_c^+$ can be defined as

$$q_c^+ = \sup\{q \geq 1, q + \varphi(q) - 1 > 0\}.$$

But, by Theorem 4 of [4], we know that for $q$ such that $f(q) > 0$, one has

$$\mathbb{E}A(t)^q < +\infty \text{ and } \sup_r \mathbb{E}A_r(t)^q < +\infty.$$

It is also known (cf. Appendix A of [22] for a proof) that, as soon as $f(2) > 0$:

$$\lim_{r \to 0}\mathbb{E}|A_r(t) - A(t)|^2 = 0.$$

The convergence of $A$ in $L^q$ for $q < q_c^+$ is a direct consequence of these two results.

Let us now give the proof of Theorem 4.1.
Let us assume that $s > t$ and $s - t > \tau$. Because $\mathbb{E}A_{(\tau,1)}^{2q}(x) < \infty$ for all $0 \leq x \leq 1$, one obtains that $\mathbb{E}A_{(\tau,1)}(t)^q A_{(\tau,1)}(s)^q < \infty$ and, using Lemma 2.1 for the 4th equality,

$$\mathbb{E}A_{(\tau,1)}{}^q(t)A_{(\tau,1)}{}^q(s)$$

$$= \mathbb{E}\left(\lim_{r_1\to 0}\int_t^{t\,|\,\tau} Q_{r_1}(x)dx\right)^q\left(\lim_{r_2\to 0}\int_s^{s+\tau} Q_{r_2}(y)dy\right)^q$$

$$= \mathbb{E}\prod_{i=1}^q \lim_{r_{1,i}\to 0}\lim_{r_{2,i}\to 0}\int_t^{t+\tau} Q_{r_{1,i}}(x_i)dx_i\int_s^{s+\tau} Q_{r_{2,i}}(y_j)dy_j$$

$$= \mathbb{E}\lim_{r\to 0}\prod_{i=1}^q\int_t^{t+\tau} Q_r(x_i)dx_i\int_s^{s+\tau} Q_r(y_i)dy_i$$

$$= \lim_{r\to 0}\mathbb{E}\prod_{i=1}^q\int_t^{t+\tau} Q_r(x_i)dx_i\int_s^{s+\tau} Q_r(y_i)dy_i$$

$$= \lim_{r\to 0}\int_{[t,t+\tau]^q}\int_{[s,s+\tau]^q}\mathbb{E}\prod_{i=1}^q Q_r(x_i)Q_r(y_i)d(x_1,...,x_q)d(y_1,...,y_q)$$

By symmetry, this yields:

$$\mathbb{E}A_{(\tau,1)}{}^q(t)A_{(\tau,1)}{}^q(s) = (q!)^2\lim_{r\to 0}\int_{D_1}\int_{D_2}\mathbb{E}\prod_{i=1}^q Q_r(x_i)Q_r(y_i)d(x_1,...,x_q)d(y_1,...,y_q).$$

where $D_1 = \{t\le x_1\le x_2,...\le x_q\le t+\tau\}$ and $D_2 = \{s\le y_1\le y_2\le ...\le y_q\le s+\tau\}$.

Let us fix $r < s-t-\tau$. Using Lemma 7.1, we can write $I = \mathbb{E}\prod_{i=1}^q Q_r(x_i)Q_r(y_i)$ for $t\le x_1\le x_2,...\le x_q\le t+\tau$ and $s\le y_1\le y_2\le ...\le y_q\le s+\tau$ as the product of 3 terms:

$$I = e^{\sum_{j=1}^q\sum_{k=1}^j\beta(j,k)\rho_r(x_k-x_j)}e^{\sum_{j=1}^q\sum_{k=1}^j\beta(j+q,k+q)\rho_r(y_k-y_j)}e^{\sum_{j=1}^q\sum_{k=1}^q\beta(j,k+q)\rho_r(y_k-x_j)}.$$
(29)

The third term (containing $\rho_r(y_k - x_j)$) controls the behavior in $|t - s|^{\varphi(2q)-2\varphi(q)}$ while the first and third terms, which do not depend on $|t-s|$, yield the multiplicative factor $|\tau|^{2(q+\varphi(q))}$. Indeed, let us consider

$$K = e^{\sum_{j=1}^q\sum_{k=1}^j\beta(j,k)\rho_r(x_k-x_j)}e^{\sum_{j=1}^q\sum_{k=1}^j\beta(j+q,k+q)\rho_r(y_k-y_j)}$$

and

$$J = e^{\sum_{j=1}^q\sum_{k=1}^q\beta(j+q,k)\rho_r(y_j,x_k)}.$$

Since $r < s-t-\tau$, $\rho_r(y_k,x_j) = -\ln|y_k - x_j|$ with $s-t-\tau\le |y_k-x_j|\le s-t+\tau$, we obtain,

$$\sum_{j=1}^q\sum_{k=1}^q\beta(j+q,k) = \sum_{j=1}^{2q}\sum_{k=1}^j\beta(j,k) - \sum_{j=1}^q\sum_{k=1}^j\beta(j,k) - \sum_{j=1}^q\sum_{k=1}^j\beta(j+q,k+q).$$

But, here, with the notations of Lemma 7.1, $\vec{p}_q = (1,1,...,1)$, $r_{k,j} = j-k+1$ and $\beta(j,k)$ depends only on $k - j$. Finally, we obtain

$$\sum_{j=1}^{q}\sum_{k=1}^{q}\beta(j+q,k)=\varphi(2q)-2\varphi(q).$$

Since $q\geq 1$ and $\varphi(2q)-2\varphi(q)<0$, we get

$$|s-t+\tau|^{\varphi(2q)-2\varphi(q)}\leq J\leq|s-t-\tau|^{\varphi(2q)-2\varphi(q)}. \tag{30}$$

Hence, by multiplying each part of 30 by K, then by taking the integral on $D_1\times D_2$ and the limit in $r=0$, we get

$$|s-t+\tau|^{\varphi(2q)-2\varphi(q)}\lim_{r\to 0}L(r,\tau)\leq$$

$$\mathbb{E}A_{(\tau,1)}{}^{q}(t)A_{(\tau,1)}{}^{q}(s)\leq|s-t-\tau|^{\varphi(2q)-2\varphi(q)}\lim_{r\to 0}L(r,\tau) \tag{31}$$

with

$$L(r,\tau)=(q!)^2\int_{D_1}\int_{D_2}e^{\sum_{j=1}^{q}\sum_{k=1}^{j}\beta(j,k)\rho_r(x_k-x_j)}e^{\sum_{j=1}^{q}\sum_{k=1}^{j}\beta(j+q,k+q)\rho_r(y_k,y_j)}.$$

It remains to show that the following convergence holds :

**Lemma 7.2.**

$$\lim_{r\to 0}L(r,\tau)=(\mathbb{E}A_{(\tau,1)}{}^{q}(t))^2=c|\tau|^{2(q+\varphi(q))}.$$

PROOF. It is known that $\mathbb{E}A_{(\tau,1)}{}^{q}(t)=c|\tau|^{q+\varphi(q)}$. However, we also have:

$$\mathbb{E}A_{(\tau,1)}{}^{q}(t)=\lim_{r\to 0}\int_{t\leq x_1,\ldots,x_q\leq t+\tau}\mathbb{E}\prod_{i=1}^{q}Q_r(x_i)d(x_1,\ldots,x_q)$$

$$=(q!)\lim_{r\to 0}\int_{D_1}\mathbb{E}\prod_{i=1}^{q}Q_r(x_i)d(x_1,\ldots,x_q)$$

which yields $\mathbb{E}A_{(\tau,1)}{}^{q}(t)\mathbb{E}A_{(\tau,1)}{}^{q}(s)=\lim_{r\to 0}L(r,\tau)$. Replacing $\lim_{r\to 0}L(r,\tau)$ with $c|\tau|^{2(q+\varphi(q))}$ in (31), we obtain

$$|s-t+\tau|^{(\varphi(2q)-2\varphi(q)}\leq\frac{\mathbb{E}A_{(\tau,1)}{}^{q}(t)A_{(\tau,1)}{}^{q}(s)}{c|\tau|^{2(q+\varphi(q))}}\leq|s-t-\tau|^{\varphi(2q)-2\varphi(q)}.$$

## 7.3 Proof of Theorem 4.2

Let us now consider the case $q>1$, $q\notin\mathbb{N}$. Let us define $q=m+\varepsilon$ with $m=[q]$ and $0<\varepsilon<1$. One writes:

$$\mathbb{E}A_{(\tau,1)}{}^{q}(s)A_{(\tau,1)}{}^{q}(t)=\mathbb{E}A_{(\tau,1)}{}^{m-1}(s)A_{(\tau,1)}{}^{m-1}(t)A_{(\tau,1)}{}^{1+\varepsilon}(s)A_{(\tau,1)}{}^{1+\varepsilon}(t).$$

(if $m = 1$, only the term $\mathbb{E}A_{(\tau,1)}^{1+\varepsilon}(s)A_{(\tau,1)}^{1+\varepsilon}(t)$ appears, but the proof is the same).

Again, $A_{(\tau,1)}^{m-1}(s)A_{(\tau,1)}^{m-1}(t)$ can be written as a multiple integral. Moreover, a classical Hölder inequality yields:

$$A_{(\tau,1)}^{1+\varepsilon}(t) = \lim_{r \to 0} \left( \int_t^{t+\tau} Q_r(x)dx \right)^{1+\varepsilon} \leq \lim_{r \to 0} \tau^\varepsilon \int_t^{t+\tau} Q_r^{1+\varepsilon}(x)dx.$$

Hence, one gets

$$\mathbb{E}A_{(\tau,1)}^q(s)A_{(\tau,1)}^q(t) \leq$$

$$\lim_{r \to 0} \tau^{2\varepsilon} \int_D \mathbb{E} \prod_{i=1}^{m-1} Q_r(x_i)Q_r(y_i)Q_r(x_m)^{1+\varepsilon}Q_r(y_m)^{1+\varepsilon}d(x_1,...,x_m)d(y_1,...,y_m)(32)$$

where $D = [t,t+\tau]^m \times [s,s+\tau]^m$. From Lemma 7.1, one can write

$$\mathbb{E} \prod_{i=1}^{m-1} Q_r(x_i)Q_r(y_i)Q_r(x_m)^{1+\varepsilon}Q_r(y_m)^{1+\varepsilon}$$

as the product of three terms. The term

$$J = e^{\sum_{j=1}^q \sum_{k=1}^q \beta(j+q,k)\rho_r(y_j,x_k)}$$

is bounded above by $|t - s + \tau|^{\varphi(2q)-2\varphi(q)}$ and the integral on $D$ of the other terms is bounded by $\tau^{2\varphi(q)+2m}$. Finally, one gets

$$\mathbb{E}A_{(\tau,1)}^q(s)A_{(\tau,1)}^q(t) \leq$$
$$\tau^{2\varepsilon}\tau^{2m+2\varphi(q)}|t-s+\tau|^{\varphi(2q)-2\varphi(q)} \leq \tau^{2q+2\varphi(2q)}|t-s+\tau|^{\varphi(2q)-2\varphi(q)}.$$

To obtain a lower bound, one writes

$$\mathbb{E}A_{(\tau,1)}^q(s)A_{(\tau,1)}^q(t) = \mathbb{E}A_{(\tau,1)}^m(s)A_{(\tau,1)}^m(t)A_{(\tau,1)}^\varepsilon(s)A_{(\tau,1)}^\varepsilon(t).$$

with

$$A_{(\tau,1)}^\varepsilon(t) = \left( \int_t^{t+\tau} Q_r(x)dx \right)^\varepsilon \geq \tau^{\varepsilon-1} \int_t^{t+\tau} Q_r^\varepsilon(x)dx.$$

With the same arguments than before, ones gets

$$\mathbb{E}A_{(\tau,1)}^q(s)A_{(\tau,1)}^q(t) \geq C(q)\tau^{2q+2\varphi(q)}|t-s+\tau|^{\varphi(2q)-2\varphi(q)}$$

where $C(q) > 0$ depends only on $q$.

### 7.4 Proof of Proposition 0.6

$$\mathbb{E}\mathscr{B}^2_{(\tau,1)}(t)\mathscr{B}^2_{(\tau,1)}(s) =$$

$$\mathbb{E}\left\{\mathbb{E}\left([B_H(u)-B_H(v)]^2[B_H(x)-B_H(y)]^2\,|\,u=A(s+\tau),v=A(s),\right.\right.$$
$$\left.\left.x=A(t+\tau),y=A(t))\right\}\right.$$

Because $X = B_H(u) - B_H(v)$ and $Y = B_H(x) - B_H(y)$ are two Gaussian vectors with finite variance, one can use the classical equality

$$\mathbb{E}X^2Y^2 = \mathbb{E}X^2\mathbb{E}Y^2 + 2(\mathbb{E}XY)^2, \qquad (33)$$

which leads by the Cauchy-Schwarz inequality to

$$\mathbb{E}X^2\mathbb{E}Y^2 \le \mathbb{E}X^2Y^2 \le 3\mathbb{E}X^2\mathbb{E}Y^2,$$

with $\mathbb{E}X^2 = \mathbb{E}(B_H(u)-B_H(v))^2 = \sigma^2|u-v|^{2H}$, $\mathbb{E}Y^2 = \sigma^2|x-y|^{2H}$.
Therefore,

$$\sigma^4\mathbb{E}A_{(\tau,1)}{}^{2H}(t)A_{(\tau,1)}{}^{2H}(s) \le \mathbb{E}\mathscr{B}^2_{(\tau,1)}(t)\mathscr{B}^2_{(\tau,1)}(s) \le 3\sigma^4\mathbb{E}A_{(\tau,1)}{}^{2H}(t)A_{(\tau,1)}{}^{2H}(s)$$

which, combined with Theorem 4.2 with $q = 2H$, gives the announced result.

## 7.5 Proof of Proposition 0.7

Because $\varphi$ is chosen such that $q_c > 4$, the quantity $I = \mathbb{E}A_{(\tau,2)}{}^2(t)A_{(\tau,2)}{}^2(s)$ is finite. Developing $A_{(\tau,2)}{}^2(t)A_{(\tau,2)}{}^2(s) = (A_{(\tau,1)}(t+\tau)-A_{(\tau,1)}(t))^2(A_{(\tau,1)}(s+\tau)-A(s))^2$ enables us to rewrite the expectation as the sum of 9 terms. Each term can be written as the integral of a product of functions $Q_r$. More precisely, with $r > t-s+2\tau$,

$$I = 4\lim_{r\to 0}\int_{u=t}^{t+\tau}\int_{v=u}^{t+\tau}\int_{x=s}^{s+\tau}\int_{y=x}^{s+\tau} J(u,v,x,y)dydxdvdu$$

with

$$J(u,v,x,y) = F(u+\tau,v+\tau,x+\tau,y+\tau)-2F(u,v+\tau,x+\tau,y+\tau)$$
$$+F(u+\tau,v+\tau,x,y)-2F(u+\tau,v+\tau,x,y+\tau)$$
$$+4F(u,v+\tau,x,y+\tau)-2F(u,v+\tau,x,y)$$
$$+F(u,v,x+\tau,y+\tau)-2F(u,v,x,y+\tau)+F(u,v,x,y)$$

where $F(a,b,c,d) = G(a,b)G(c,d)H(a,b,c,d)$ for $a \le b \le c \le d$ with

$$G(z_1,z_2) = G_r(z_1,z_2) = e^{\beta(2,1)\rho_r(z_2-z_1)}$$

and

$$H(a,b,c,d) = e^{-\beta(3,2)\rho_r(c-b)}e^{\beta(3,1)\rho_r(c-a)}e^{\beta(4,2)\rho_r(d-b)}e^{\beta(1,4)\rho_r(d-a)}$$
$$= (c-b)^{\beta(3,2)}(c-a)^{\beta(3,1)}(d-b)^{\beta(4,2)}(d-a)^{\beta(4,1)}.$$

Hence, $J$ can be written as

$$J = -\Delta_x\Delta_v F(u,v,x,y+\tau) + \Delta_u\Delta_x F(u,v+\tau,x,y+\tau)$$
$$+\Delta_v\Delta_y F(u,v,x,y) - \Delta_u\Delta_y F(u,v+\tau,x,y)$$

where $\Delta_u F(u,v,x,y)$ (resp. $\Delta_v$, $\Delta_x$ and $\Delta_y$) denotes the forward difference $F(u+\tau,v,x,y) - F(u,v,x,y)$ (resp. $F(u,v+\tau,x,y) - F(u,v,x,y)$, etc.).

Expressing $J$ in terms of $G$ and $H$, one finds

$$J = \Delta_u G(u,v+\tau)\Delta_x G(x,y+\tau)H(u,v+\tau,x,y+\tau)$$
$$+G(u,v+\tau)\Delta_x G(x,y+\tau)\Delta_u H(u,v+\tau,x,y+\tau)$$
$$+\Delta_u G(u,v+\tau)G(x,y+\tau)\Delta_x H(u,v+\tau,x,y+\tau)$$
$$+G(u,v+\tau)G(x,y+\tau)\Delta_u\Delta_x H(u,v+\tau,x,y+\tau)$$
$$-\Delta_x G(x,y+\tau)\Delta_v G(u,v)H(u,v,x,y+\tau)$$
$$-G(x,y+\tau)\Delta_v G(u,v)\Delta_x H(u,v,x,y+\tau)$$
$$-\Delta_x G(x,y+\tau)G(u,v)\Delta_v H(u,v,x,y+\tau)$$
$$-G(x,y+\tau)G(u,v)\Delta_x\Delta_v H(u,v,x,y+\tau)$$
$$+\Delta_y G(x,y)\Delta_v G(u,v)H(u,v,x,y)$$
$$+\Delta_y G(x,y)G(u,v)\Delta_v H(u,v,x,y)$$
$$+G(x,y)\Delta_v G(u,v)\Delta_y H(u,v,x,y)$$
$$+G(x,y)G(u,v)\Delta_v\Delta_y H(u,v,x,y)$$
$$-\Delta_y G(x,y)\Delta_u G(u,v+\tau)H(u,v+\tau,x,y)$$
$$-\Delta_y G(x,y)G(u,v+\tau)\Delta_u H(u,v+\tau,x,y)$$
$$-G(x,y)\Delta_u G(u,v+\tau)\Delta_y H(u,v+\tau,x,y)$$
$$-G(x,y)G(u,v+\tau)\Delta_u\Delta_v H(u,v+\tau,x,y).$$

We first consider the terms where one or two finite differences are taken on $H$, for example, $G(x,y+\tau)\Delta_v G(u,v)\Delta_x H(u,v,x,y+\tau)$. One has, for any $0 < r < s-t-2\tau$,

$$|\Delta_x H(u,v,x,y+\tau)| \leq \tau|t-s-2\tau|^{\varphi(4)-2\varphi(2)-1} + \tau R(\tau),$$

with $R(\tau) \to 0$ when $\tau \to 0$, $\Delta_v G(u,v) \leq 0$ and $G(x,y+\tau) \geq 0$. Therefore,

$$A = \int\int\int\int G(x,y+\tau)\Delta_v G(u,v)\Delta_x H(u,v,x,y+\tau)dydxdvdu$$
$$\leq \tau|t-s-2\tau|^{\varphi(4)-2\varphi(2)-1}\left(\int\int G(x,y+\tau))dxdy\right)\left(\int\int \Delta_v G(u,v)dudv\right).$$

Besides, we have the following result whose proof is postponed to the end of this subsection.

**Lemma 7.3.** *There exists $c_1 > 0$ and $c_2 > 0$ such that*

$$\lim_{r \to 0} \int_{x=s}^{s+\tau} \int_{y=x}^{s+\tau} G(x, y+\tau) dy dx \le \mathbb{E}A_{(2\tau,1)}^2(t) = c_1 |\tau|^{\varphi(2)+2}$$

*and*

$$\lim_{r \to 0} \int_{u=t}^{t+\tau} \int_{v=u}^{t+\tau} \Delta_u G(u, v+\tau) dv du = \mathbb{E}A_{(\tau,2)}^2(t)/4 = c_2 |\tau|^{\varphi(2)+2}.$$

So,

$$\lim_{\tau \to 0} \frac{\int\int\int\int G(x, y+\tau)\Delta_v G(u, v)\Delta_v H(u, v, x, y+\tau) du dv dx dy}{|\tau|^{2(2+\varphi(2))}} = 0.$$

The other terms with finite difference on $H$ can be dealt with in a similar way. Finally, since

$$\lim_{r \to 0} \Delta_u G(u, v+\tau) =$$
$$- \lim_{r \to 0} \Delta_v G(u, v) = \lim_{r \to 0} \Delta_x G(x, y+\tau) = - \lim_{r \to 0} \Delta_x G(x, y) = 1/4 \mathbb{E}A_{(\tau,2)}^2(t),$$

and since

$$|s - t + 2\tau|^{\varphi(4)-2\varphi(2)} \le H(u, v+\tau, x, y+\tau) \le |s - t - 2\tau|^{\varphi(4)-2\varphi(2)},$$

$$|s - t + 2\tau|^{\varphi(4)-2\varphi(2)} \le H(u, v+\tau, x, y) \le |s - t - 2\tau|^{\varphi(4)-2\varphi(2)},$$

we obtain the announced result.

Let us now give the proof of Lemma 7.3. The first inequality is trivial since we have

$$\mathbb{E}A_{(2\tau,1)}^2(t) = \lim_{r \to 0} \int_t^{t+2\tau} \int_t^{t+2\tau} \mathbb{E}Q_r(x)Q_r(y) dx dy$$
$$\ge \int_t^{t+\tau} \int_t^{t+\tau} \mathbb{E}Q_r(x)Q_r(y+\tau) dx dy$$
$$= \int_t^{t+\tau} \int_t^{t+\tau} G(x, y+\tau) dx dy.$$

For the second point, let us remark first, that

$$\mathbb{E}A_{(\tau,2)}^2(t) = (A_{(\tau,1)}(t+\tau) - A_{(\tau,1)}(t))^2$$
$$= A_{(\tau,1)}(t+\tau)^2 - 2A_{(\tau,1)}(t)A_{(\tau,1)}(t+\tau) + A_{(\tau,1)}^2(t)$$
$$= \lim_{r \to 0} \int_t^{t+\tau} \int_t^{t+\tau} (\mathbb{E}Q_r(x+\tau)Q_r(y+\tau) - 2\mathbb{E}Q_r(x)Q_r(y+\tau)$$
$$+ \mathbb{E}Q_r(x)Q_r(x)) dx dy).$$

But, $\mathbb{E}Q_r(x+\tau)Q_r(y+\tau) = \mathbb{E}Q_r(x)Q_r(y)$ and

$$\mathbb{E}A_{(\tau,2)}{}^2(s) = 2\lim_{r\to 0}\int_t^{t+\tau}\int_t^{t+\tau}(\mathbb{E}Q_r(x+\tau)Q_r(y+\tau) - \mathbb{E}Q_r(x)Q_r(y+\tau))dxdy$$

$$= 4\lim_{r\to 0}\int_{x=s}^{t+\tau}\int_{y=x}^{t+\tau}\Delta_x G(x,y+\tau)dxdy.$$

We still have to show that $\lim_{r\to 0}\int\int\Delta_u G(u,v+\tau)dudv = c\tau^{\varphi(2)+2}$. First, note that

$$\lim_{r\to 0}\int_{u=t}^{t+\tau}\int_{v=u}^{t+\tau}\Delta_u G(u,v+\tau)dudv =$$

$$\lim_{r\to 0}\int_{u=t}^{t+\tau}\int_{v=u}^{t+\tau}(G(u+\tau,v+\tau) - G(u,v+\tau))dudv.$$

But,

$$\lim_{r\to 0}\int_{u=t}^{t+\tau}\int_{v=u}^{t+\tau}G(u+\tau,v+\tau)dudv = 1/2\lim_{r\to 0}\int_{u=t}^{t+\tau}\int_{v=t}^{t+\tau}\mathbb{E}Q_r(u)Q_r(v)dudv$$

$$= 1/2\mathbb{E}A_{(\tau,1)}{}^2(t) = c|\tau|^{\varphi(2)+2}$$

for some $c > 0$ and

$$\lim_{r\to 0}\int_{u=t}^{t+\tau}\int_{v=u}^{t+\tau}G(u,v+\tau) = \lim_{r\to 0}\int_{u=t}^{t+\tau}\int_{v=u}^{t+\tau}|v+\tau-u|^{\varphi(2)}dvdu = \tilde{c}|\tau|^{\varphi(2)+2}.$$

for some $\tilde{c} > 0$. Thus,

$$\lim_{r\to 0}\int_{u=t}^{t+\tau}\int_{v=u}^{t+\tau}\Delta_u G(u,v+\tau)dvdu = C|\tau|^{\varphi(2)+2}$$

for a $C \in \mathbb{R}$. Since $\lim_{r\to 0}\int\int\Delta_u G(u,v)dudv = \mathbb{E}A_{(\tau,2)}{}^2(s)/4 > 0$, it comes that $C > 0$ and Lemma 7.3 is proven.

**Remark 2:** The crucial point in the above proof consists of the fact that when $q \geq 2$, the covariance function can be split into a number of terms, expressed with auto-terms $G$ and cross terms $H$ (whose that involve the dependence in $|t - s|$). Increments - of any order $P$ - for some of these terms apply only to the auto-terms $G$ and hence do not produce in reduction the in the rate of decrease in $|t - s|$ (only related to the cross terms $H$). A contrario, for $q = 1$, $G \equiv 1$ and the increments are taken on $H$, leading to Formula (22). This qualitative argument indicates that $P$ does not play any role in the control of higher order correlation functions and is founding for the formulation of Conjectures 0.1 and 0.2 in Section 5.

**Remark 3:** Also, it is worth mentioning that in this derivation of the proof of Proposition 0.6, one obtains a number of higher order terms in the expansion in $|\tau|$, whose impact is significant, so that the leading power law term may be difficult to observe practically (cf. Discussion in Section 5.2.3).

## 7.6 Proof of Proposition 0.8

$$\mathbb{E}\mathscr{B}^2_{(\tau,2)}(t)\mathscr{B}^2_{(\tau,2)}(s)$$
$$= \mathbb{E}\left(\mathbb{E}[B_H(w) - 2B_H(v) + B_H(u))]^2 [B_H(z) - 2B_H(y) + B_H(x)]^2\right|$$
$$w = A(s+2\tau), v = A(s+\tau), u = A(s), z = A(t+2\tau), y = A(t+\tau), x = A(t))$$

where $X = B_H(w) - 2B_H(v) + B_H(u)$ and $Y = B_H(z) - 2B_H(y) + B_H(x)$ are two Gaussian vectors with finite variance. Equality (33) and Cauchy-Schwarz inequality yield

$$\mathbb{E}X^2\mathbb{E}Y^2 \le \mathbb{E}X^2Y^2 \le 3\mathbb{E}X^2\mathbb{E}Y^2$$

with

$$\mathbb{E}X^2 = \mathbb{E}(B_H(w) - 2B_H(v) + B_H(u))^2 = \sigma^2(2|w-v|^{2H} - |w-u|^{2H} + 2|v-u|^{2H}).$$

In addition, since $A$ is a non decreasing function,

$$2|A(t+2\tau) - A(t+\tau)|^{2H} - |A(t+2\tau) - A(t)|^{2H} + 2|A(t+\tau) - A(t)|^{2H}$$
$$\le 2A_{(\tau,1)}^{2H}(t+\tau) + 2A_{(\tau,1)}^{2H}(t) \le 2A_{(2\tau,1)}^{2H}(t)$$

and there exists $C_H > 0$ such that

$$2|A(t+2\tau) - A(t+\tau)|^{2H} + |A(t+2\tau) - A(t)|^{2H} + 2|A(t+\tau) - A(t)|^{2H}$$
$$\ge C_H(A_{(\tau,1)}(t+\tau) + A_{(\tau,1)}(t)) \ge C_H A_{(\tau,1)}(t) \quad.$$

Finally, it comes

$$C_H\sigma^4\mathbb{E}A_{(\tau,1)}^{2H}(t)A_{(\tau,1)}^{2H}(s) \le \mathbb{E}\mathscr{B}^2_{(\tau,2)}(t)\mathscr{B}^2_{(\tau,2)}(s) \le 4\sigma^4\mathbb{E}A_{(2\tau,1)}^{2H}(t)A_{(2\tau,1)}^{2H}(s),$$

which, combined with Theorem 4.2, leads to the result.

**Acknowledgements** This work has mostly been completed while B. Vedel was at ENS Lyon, on a Post-Doctoral Grant supported by the gratefully acknowledged *Foundation Del Duca, Institut de France, 2007*. The authors gratefully acknowledge an anonymous reviewer for numerous and detailed valuable technical comments

## References

[1] P. Abry, P. Flandrin, M. Taqqu, and D. Veitch. Wavelets for the analysis, estimation, and synthesis of scaling data. In K. Park and W. Willinger, editors, *Self-similar Network Traffic and Performance Evaluation*, pages 39–88, New York, 2000. Wiley.

 [2] P. Abry, P. Gonçalvès, and P. Flandrin. *Wavelets, spectrum estimation and 1/f processes*, chapter 103. Springer-Verlag, New-York, 1995. Wavelets and Statistics, Lecture Notes in Statistics.

 [3] A. Arneodo, E. Bacry, and J.F. Muzy. Random cascades on wavelet dyadic trees. *J. Math. Phys.*, 39(8):4142–4164, 1998.

 [4] E. Bacry and J.F. Muzy. Log-infinitely divisible multifractal processes. *Commun. Math. Phys.*, 236:449–475, 2003.

 [5] J. Barral and B. Mandelbrot. Multiplicative products of cylindrical pulses. *Probab. Theory Rel.*, 124:409–430, 2002.

 [6] J. Beran. *Statistics for Long-Memory Processes*. Chapman & Hall, 1994.

 [7] P. Chainais. *Cascades log-infiniment divisibles et analyse multirésolution. Application à l'étude des intermittences en turbulence*. PhD thesis, Ecole Normale Supérieure de Lyon, 2001.

 [8] P. Chainais, R. Riedi, and P. Abry. On non scale invariant infinitely divisible cascades. *IEEE Trans. Inform. Theory*, 51(3), March 2005.

 [9] P. Chainais, R. Riedi, and P. Abry. Warped infinitely divisible cascades: beyond scale invariance. *Traitement du Signal*, 22(1), 2005.

 [10] J.-F. Coeurjolly. Estimating the parameters of a fractional Brownian motion by discrete variations of its sample paths. *Statistical Inference for Stochastic Processes*, 4(2):199–227, 2001.

 [11] William Feller. *An Introduction to Probability Theory and Its Applications*, volume 2. John Wiley and Sons, Inc., New-York, London, Sidney, 1966.

 [12] P. Flandrin. On the spectrum of fractional Brownian motions. *IEEE Trans. on Info. Theory*, IT-35(1):197–199, 1989.

 [13] P. Flandrin. Wavelet analysis and synthesis of fractional Brownian motions. *IEEE Trans. Inform. Theory*, 38:910–917, 1992.

 [14] U. Frisch. *Turbulence, the Legacy of A.N. Kolmogorov*. Cambridge University Press, 1995.

 [15] P. Gonçalvès and R. Riedi. Wavelet analysis of fractional Brownian motion in multifractal time. In *Proc. GRETSI Symposium Signal and Image Processing*, Vannes, France, 1999.

 [16] J. Istas and G. Lang. Quadratic variations and estimation of the local Hölder index of a Gaussian process. *Annales de l'institut Henri Poincaré (B) Probabilités et Statistiques*, 33(4):407–436, 1997.

 [17] S. Jaffard. Wavelet techniques in multifractal analysis. In M. Lapidus and M. van Frankenhuijsen, editors, *Fractal Geometry and Applications: A Jubilee of Benoît Mandelbrot, M. Lapidus and M. van Frankenhuijsen Eds., Proceedings of Symposia in Pure Mathematics*, volume 72(2), pages 91–152. AMS, 2004.

 [18] S. Jaffard, B. Lashermes, and P. Abry. Wavelet leaders in multifractal analysis. In T. Qian, M.I. Vai, and X. Yuesheng, editors, *Wavelet Analysis and Applications*, pages 219–264, Basel, Switzerland, 2006. Birkhäuser Verlag.

 [19] B. Lashermes, P. Abry, and P. Chainais. New insight in the estimation of scaling exponents. *Int. J. Wavelets Multi.*, 2(4):497–523, 2004.

[20] B.B. Mandelbrot. Intermittent turbulence in self-similar cascades: divergence of high moments and dimension of the carrier. *J. Fluid Mech.*, 62:331–358, 1974.

[21] B.B. Mandelbrot. A multifractal walk down Wall Street. *Sci. Am.*, 280(2):70–73, 1999.

[22] J.F. Muzy and E. Bacry. Multifractal stationary random measures and multifractal random walks with log-infinitely divisible scaling laws. *Phys. Rev. E*, 66, 2002.

[23] M. Ossiander and E.C. Waymire. Statistical estimation for multiplicative cascades. *Ann. Stat.*, 28(6):1533–1560, 2000.

[24] G. Samorodnitsky and M. Taqqu. *Stable non-Gaussian random processes.* Chapman and Hall, New York, 1994.

[25] F. Schmitt and D. Marsan. Stochastic equations generating continuous multiplicative cascades. *Eur. Phys. J. B*, 20(1):3–6, 2001.

[26] M. S. Taqqu. Fractional brownian motion and long-range dependence. In P. Doukhan, G. Oppenheim, and M. S. Taqqu, editors, *Theory and Applications of Long-Range Dependence*, pages 5–38. Birkhäuser, 2002.

[27] A.H. Tewfik and M. Kim. Correlation structure of the discrete wavelet coefficients of fractional Brownian motions. *IEEE Trans. on Info. Theory*, IT-38(2):904–909, 1992.

[28] D. Veitch and P. Abry. A wavelet-based joint estimator of the parameters of long-range dependence. *IEEE Trans. Inform. Theory*, 45(3):878–897, 1999.

[29] H. Wendt. *Contributions of Wavelet Leaders and Bootstrap to Multifractal Analysis: Images, Estimation Performance, Dependence Structure and Vanishing Moments. Confidence Intervals and Hypothesis Tests.* PhD thesis, Ecole Normale Supérieure de Lyon, France, 2008.

[30] H. Wendt and P. Abry. Multifractality tests using bootstrapped wavelet leaders. *IEEE Trans. Signal Proces.*, 55(10):4811–4820, 2007.

[31] H. Wendt, P. Abry, and S.Jaffard. Bootstrap for empirical multifractal analysis. *IEEE Signal Proc. Mag.*, 24(4):38–48, 2007.

[32] G.W. Wornell. A Karhunen-Loève-like expansion for $1/f$ processes *via* wavelets. *IEEE Trans. on Info. Theory*, IT-36(4):859–861, 1990.

# Multifractal scenarios for products of geometric Ornstein-Uhlenbeck type processes

Vo V. Anh, Nikolai N. Leonenko, and Narn-Rueih Shieh

**Abstract** We investigate the properties of multifractal products of the exponential of Ornstein-Uhlenbeck processes driven by Lévy motion. The conditions on the mean, variance and covariance functions of these processes are interpreted in terms of the moment generating functions. We provide two examples of tempered stable and normal tempered stable distributions. We establish the corresponding log-tempered stable and log-normal tempered stable scenarios, including their Rényi functions and dependence structures.

## 1 Introduction

Multifractal models have been used in many applications in hydrodynamic turbulence, finance, genomics, computer network traffic, etc. (see, for example, Kolmogorov 1941, 1962, Kahane 1985, 1987, Gupta and Waymire 1993, Novikov 1994, Frisch 1995, Mandelbrot 1997, Falconer 1997, Schertzer *et al.* 1997, Anh *et al.* 2001, Abry *et al.* 2002, Hohn *et al.* 2003, Yang *et al.* 2008). There are many ways to construct random multifractal measures ranging from simple binomial cascades

Vo V. Anh
School of Mathematical Sciences,
Queensland University of Technology
GPO Box 2434, Brisbane QLD 4001, Australia, e-mail: v.anh@qut.edu.au

Nikolai N. Leonenko
Cardiff School of Mathematics
Cardiff University
Senghennydd Road, Cardiff CF24 4YH, UK, e-mail: LeonenkoN@Cardiff.ac.uk

Narn-Rueih Shieh
Department of Mathematics
National Taiwan University
Taipei 10617, Taiwan, e-mail: shiehnr@math.ntu.edu.tw

P. Doukhan et al. (eds.), *Dependence in Probability and Statistics*,
Lecture Notes in Statistics 200, DOI 10.1007/978-3-642-14104-1_6,
© Springer-Verlag Berlin Heidelberg 2010

to measures generated by branching processes and the compound Poisson process (Kahane 1985, 1987, Gupta and Waymire 1993, Falconer 1997, Schmitt and Marsan 2001, Barral and Mandelbrot 2002, Bacry and Muzy 2003, Riedi 2003, Mörters and Shieh 2002, 2004, 2008, Shieh and Taylor 2002, Schmitt 2003, Schertzer and Lovejoy 2004, Chainais *et al.* 2005). Jaffard (1999) showed that Lévy processes (except Brownian motion and Poisson processes) are multifractal; but since the increments of a Lévy process are independent, this class excludes the effects of dependence structures. Moreover, Lévy processes have a linear singularity spectrum while real data often exhibit a strictly concave spectrum.

Anh, Leonenko and Shieh (2008, 2009a,b) considered multifractal products of stochastic processes as defined in Kahane (1985, 1987) and Mannersalo *et al.* (2002). Especially Anh, Leonenko and Shieh (2008) constructed multifractal processes based on products of geometric Ornstein-Uhlenbeck processes driven by Lévy motion with inverse Gaussian or normal inverse Gaussian distribution. They also described the behaviour of the $q$-th order moments and Rényi functions, which are nonlinear, hence displaying the multifractality of the processes as constructed.

This paper constructs two new multifractal processes which generalize those corresponding to the inverse Gaussian and normal inverse Gaussian distributions obtained in Anh, Leonenko and Shieh (2008). We use the theory of Ornstein-Uhlenbeck processes with tempered stable law and normal tempered stable law for their marginal distributions (see Barndorff-Nielsen and Shephard 2002, Terdik and Woyczynski 2004, and the references therein). Note that the tempered stable distribution is also known (up to constants) as the Vershik-Yor subordinator (see Donati-Martin and Yor 2006, and the references therein).

The next section recalls some basic results on multifractal products of stochastic processes as developed in Kahane (1985, 1987) and Mannersalo *et al.* (2002). Section 3 recalls some properties of Lévy processes and Ornstein-Uhlenbeck processes as strong solutions of stochastic differential equations (of the form (15) below) driven by a Lévy process. These properties are needed to construct multifractal Ornstein-Uhlenbeck processes in Section 4. In particular we demonstrate the construction for the cases when the background driving Lévy process has the tempered stable distribution and the normal tempered stable distribution.

## 2 Multifractal products of stochastic processes

This section recaptures some basic results on multifractal products of stochastic processes as developed in Kahane (1985, 1987) and Mannersalo *et al.* (2002). We provide a new interpretation of their conditions based on the moment generating functions, which is useful for our exposition.

We introduce the following conditions:

**A′.** Let $\Lambda(t)$, $t \in \mathbb{R}_+ = [0, \infty)$, be a measurable, separable, strictly stationary, positive stochastic process with $E\Lambda(t) = 1$.

We call this process the mother process and consider the following setting:
**A″**. Let $\Lambda^{(i)}$, $i = 0, 1, \ldots$ be independent copies of the mother process $\Lambda$, and $\Lambda_b^{(i)}$ be the rescaled version of $\Lambda^{(i)}$ :

$$\Lambda_b^{(i)}(t) \stackrel{d}{=} \Lambda^{(i)}(tb^i), \quad t \in \mathbb{R}_+, \quad i = 0, 1, 2, \ldots,$$

where the scaling parameter $b > 1$, and $\stackrel{d}{=}$ denotes equality in finite-dimensional distributions.

Moreover, in the examples of Section 4, the stationary mother process satisfies the following conditions:
**A‴**. For $t \in \mathbb{R}_+$, let $\Lambda(t) = \exp\{X(t)\}$, where $X(t)$ is a stationary process with $EX^2(t) < \infty$.

We denote by $\theta \in \Theta \subseteq \mathbb{R}^p, p \geqslant 1$ the parameter vector of the distribution of the process $X(t)$ and assume that there exist a marginal probability density function $p_\theta(x)$ and a bivariate probability density function $p_\theta(x_1, x_2; t_1 - t_2)$ such that the moment generating function

$$M(\zeta) = E\exp\{\zeta X(t)\}$$

and the bivariate moment generating function

$$M(\zeta_1, \zeta_2; t_1 - t_2) = E\exp\{\zeta_1 X(t_1) + \zeta_2 X(t_2)\}$$

exist.

The conditions **A′-A‴** yield

$$E\Lambda_b^{(i)}(t) = M(1) = 1;$$
$$\mathrm{Var}\Lambda_b^{(i)}(t) = M(2) - 1 = \sigma_\Lambda^2 < \infty;$$

$$\mathrm{Cov}(\Lambda_b^{(i)}(t_1), \Lambda_b^{(i)}(t_2)) = M(1, 1; (t_1 - t_2)b^i) - 1, \ b > 1.$$

We define the finite product processes

$$\Lambda_n(t) = \prod_{i=0}^n \Lambda_b^{(i)}(t) = \exp\left\{\sum_{i=0}^n X^{(i)}(tb^i)\right\}, \tag{1}$$

and the cumulative processes

$$A_n(t) = \int_0^t \Lambda_n(s)ds, \quad n = 0, 1, 2, \ldots, \tag{2}$$

where $X^{(i)}(t), i = 0, \ldots, n, \ldots$, are independent copies of the stationary process $X(t), t \geq 0$.

We also consider the corresponding positive random measures defined on Borel sets $B$ of $\mathbb{R}_+$ :

$$\mu_n(B) = \int_B \Lambda_n(s)ds, \quad n = 0, 1, 2, \ldots \tag{3}$$

Kahane (1987) proved that the sequence of random measures $\mu_n$ converges weakly almost surely to a random measure $\mu$. Moreover, given a finite or countable family of Borel sets $B_j$ on $\mathbb{R}_+$, it holds that $\lim_{n\to\infty} \mu_n(B_j) = \mu(B_j)$ for all $j$ with probability one. The almost sure convergence of $A_n(t)$ in countably many points of $\mathbb{R}_+$ can be extended to all points in $\mathbb{R}_+$ if the limit process $A(t)$ is almost surely continuous. In this case, $\lim_{n\to\infty} A_n(t) = A(t)$ with probability one for all $t \in \mathbb{R}_+$. As noted in Kahane (1987), there are two extreme cases: (i) $A_n(t) \to A(t)$ in $L_1$ for each given $t$, in which case $A(t)$ is not almost surely zero and is said to be fully active (non-degenerate) on $\mathbb{R}_+$; (ii) $A_n(1)$ converges to 0 almost surely, in which case $A(t)$ is said to be degenerate on $\mathbb{R}_+$. Sufficient conditions for non-degeneracy and degeneracy in a general situation and relevant examples are provided in Kahane (1987) (Eqs. (18) and (19) respectively.) The condition for complete degeneracy is detailed in Theorem 3 of Kahane (1987).

The Rényi function, also known as the deterministic partition function, is defined for $t \in [0, 1]$ as

$$T(q) = \liminf_{n\to\infty} \frac{\log E \sum_{k=0}^{2^n-1} \mu^q \left( I_k^{(n)} \right)}{\log \left| I_k^{(n)} \right|}$$

$$= \liminf_{n\to\infty} \left( -\frac{1}{n} \right) \log_2 E \sum_{k=0}^{2^n-1} \mu^q \left( I_k^{(n)} \right), \tag{4}$$

where $I_k^{(n)} = [k2^{-n}, (k+1)2^{-n}]$, $k = 0, 1, \ldots, 2^n - 1$, $\left| I_k^{(n)} \right|$ is its length, and $\log_b$ is log to the base $b$.

*Remark 0.1.* The multifractal formalism for random cascades can be stated in terms of the Legendre transform of the Rényi function:

$$T^*(\alpha) = \min_{q \in \mathbb{R}} (q\alpha - T(q)).$$

In fact, let $f(\alpha)$ be the Hausdorff dimension of the set

$$C_\alpha = \left\{ t \in [0, 1] : \lim_{n\to\infty} \frac{\log \mu \left( I_k^{(n)}(t) \right)}{\log \left| I_k^{(n)} \right|} = \alpha \right\},$$

where $I_k^{(n)}(t)$ is a sequence of intervals $I_k^{(n)}$ that contain $t$. The function $f(\alpha)$ is known as the singularity spectrum of the measure $\mu$, and we refer to $\mu$ as a multifractal measure if $f(\alpha) \neq 0$ for a continuum of $\alpha$. When the relationship

$$f(\alpha) = T^*(\alpha) \tag{5}$$

is established for a measure $\mu$, we say that the multifractal formalism holds for this measure.

Mannersalo *et al.* (2002) presented the conditions for $L_2$-convergence and scaling of moments:

**Theorem 2.1.** *(Mannersalo, Norros and Riedi 2002) Suppose that the conditions $A'$-$A'''$ hold.*

*If, for some positive numbers $\delta$ and $\gamma$,*

$$\exp\{-\delta |\tau|\} \leqslant \rho(\tau) = \frac{M(1,1;\tau)-1}{M(2)-1} \leqslant |C\tau|^{-\gamma}, \tag{6}$$

*then $A_n(t)$ converges in $L_2$ if and only if*

$$b > 1 + \sigma_\Lambda^2 = M(2).$$

*If $A_n(t)$ converges in $L_2$, then the limit process $A(t)$ satisfies the recursion*

$$A(t) = \frac{1}{b}\int_0^t \Lambda(s)d\widetilde{A}(bs), \tag{7}$$

*where the processes $\Lambda(t)$ and $\widetilde{A}(t)$ are independent, and the processes $A(t)$ and $\widetilde{A}(t)$ have identical finite-dimensional distributions.*

*If $A(t)$ is non-degenerate, the recursion (7) holds, $A(1) \in L_q$ for some $q > 0$, and $\sum_{n=0}^\infty c(q,b^{-n}) < \infty$, where $c(q,t) = \mathrm{E}\sup_{s \in [0,t]} |\Lambda^q(0) - \Lambda^q(s)|$, then there exist constants $\overline{C}$ and $\underline{C}$ such that*

$$\underline{C}t^{q-\log_b \mathrm{E}\Lambda^q(t)} \leqslant \mathrm{E}A^q(t) \leqslant \overline{C}t^{q-\log_b \mathrm{E}\Lambda^q(t)}, \tag{8}$$

*which will be written as*

$$\mathrm{E}A^q(t) \sim t^{q-\log_b \mathrm{E}\Lambda^q(t)}, \ t \in [0,1].$$

*If, on the other hand, $A(1) \in L_q$, $q > 1$, then the Rényi function is given by*

$$T(q) = q - 1 - \log_b \mathrm{E}\Lambda^q(t) = q - 1 - \log_b M(q). \tag{9}$$

*If $A(t)$ is non-degenerate, $A(1) \in L_2$, and $\Lambda(t)$ is positively correlated, then*

$$\mathrm{Var}A(t) \geqslant \mathrm{Var}\int_0^t \Lambda(s)ds. \tag{10}$$

Hence, if $\int_0^t \Lambda(s)ds$ is strongly dependent, then $A(t)$ is also strongly dependent.

*Remark 0.2. The result (8) means that the process $A(t)$, $t \in [0,1]$ with stationary increments behaves as*

$$\log \mathrm{E}\left[A(t+\delta)-A(t)\right]^{q} \sim K(q)\log\delta + C_{q} \tag{11}$$

*for a wide range of resolutions $\delta$ with a nonlinear function*

$$K(q) = q - \log_{b}\mathrm{E}\Lambda^{q}(t) = q - \log_{b}M(q),$$

*where $C_{q}$ is a constant. In this sense, the stochastic process $A(t)$ is said to be multifractal. The function $K(q)$, which contains the scaling parameter $b$ and all the parameters of the marginal distribution of the mother process $X(t)$, can be estimated by running the regression (11) for a range of values of $q$. For the example in Section 4, the explicit form of $K(q)$ is obtained. Hence these parameters can be estimated by minimizing the mean square error between the $K(q)$ curve estimated from data and its analytical form for a range of values of $q$. This method has been used for multifractal characterization of complete genomes in Anh et al. (2001).*

## 3 Geometric Ornstein-Uhlenbeck processes

This section reviews a number of known results on Lévy processes (see Skorokhod 1991, Bertoin 1996, Sato 1999, Kyprianou 2006) and Ornstein-Uhlenbeck type processes (see Barndorff-Nielsen 2001, Barndorff-Nielsen and Shephard 2001). As standard notation we will write

$$\kappa(z) = C\{z;X\} = \log \mathrm{E}\exp\{izX\}, \quad z \in \mathbb{R}$$

for the cumulant function of a random variable $X$, and

$$K\{\zeta;X\} = \log \mathrm{E}\exp\{\zeta X\}, \quad \zeta \in \mathbb{R}$$

for the Lévy exponent or Laplace transform or cumulant generating function of the random variable $X$. Its domain includes the imaginary axis and frequently larger areas.

A random variable $X$ is *infinitely divisible* if its cumulant function has the Lévy-Khintchine form

$$C\{z;X\} = iaz - \frac{d}{2}z^{2} + \int_{\mathbb{R}}\left(e^{izu} - 1 - izu\mathbf{1}_{[-1,1]}(u)\right)\nu(du), \tag{12}$$

where $a \in \mathbb{R}$, $d \geq 0$ and $\nu$ is the Lévy measure, that is, a non-negative measure on $\mathbb{R}$ such that

$$\nu(\{0\}) = 0, \quad \int_{\mathbb{R}}\min\left(1,u^{2}\right)\nu(du) < \infty. \tag{13}$$

The triplet $(a,d,\nu)$ uniquely determines the random variable $X$. For a Gaussian random variable $X \sim N(a,d)$, the Lévy triplet takes the form $(a,d,0)$.

A random variable $X$ is *self-decomposable* if, for all $c \in (0,1)$, the characteristic function $f(z)$ of $X$ can be factorized as $f(z) = f(cz)f_{c}(z)$ for some characteristic function $f_c(z)$, $z \in \mathbb{R}$. A homogeneous Lévy process $Z = \{Z(t), t \geq 0\}$ is a continuous (in probability), càdlàg process with independent and stationary increments and $Z(0) = 0$ (recalling that a càdlàg process has right-continuous sample paths with existing left limits.) For such processes we have $C\{z; Z(t)\} = tC\{z; Z(1)\}$ and $Z(1)$ has the Lévy-Khintchine representation (12).

Let $f(z)$ be the characteristic function of a random variable $X$. If $X$ is self-decomposable, then there exists a stationary stochastic process $\{X(t), t \geq 0\}$, such that $X(t) \stackrel{d}{=} X$ and

$$X(t) = e^{-\lambda t} X(0) + \int_{(0,t]} e^{-\lambda(t-s)} dZ(\lambda s) \tag{14}$$

for all $\lambda > 0$ (see Barndorff-Nielsen 1998). Conversely, if $\{X(t), t \geq 0\}$ is a stationary process and $\{Z(t), t \geq 0\}$ is a Lévy process, independent of $X(0)$, such that $X(t)$ and $Z(t)$ satisfy the Itô stochastic differential equation

$$dX(t) = -\lambda X(t) dt + dZ(\lambda t) \tag{15}$$

for all $\lambda > 0$, then $X(t)$ is self-decomposable. A stationary process $X(t)$ of this kind is said to be an Ornstein-Uhlenbeck type process or an OU-type process, for short. The process $Z(t)$ is termed the background driving Lévy process (BDLP) corresponding to the process $X(t)$. In fact (14) is the unique (up to indistinguishability) strong solution to Eq. (15) (Sato 1999, Section 17). The meaning of the stochastic integral in (14) was detailed in Applebaum (2004, p. 214).

Let $X(t)$ be a square integrable OU process. Then $X(t)$ has the correlation function

$$r_X(t) = \exp\{-\lambda |t|\}, \quad t \in \mathbb{R}. \tag{16}$$

The cumulant transforms of $X = X(t)$ and $Z(1)$ are related by

$$C\{z; X\} = \int_0^\infty C\{e^{-s}z; Z(1)\} ds = \int_0^z C\{\xi; Z(1)\} \frac{d\xi}{\xi}$$

and

$$C\{z; Z(1)\} = z \frac{\partial C\{z; X\}}{\partial z}.$$

Suppose that the Lévy measure $v$ of $X$ has a density function $p(u)$, $u \in \mathbb{R}$, which is differentiable. Then the Lévy measure $\tilde{v}$ of $Z(1)$ has a density function $q(u)$, $u \in \mathbb{R}$, and $p$ and $q$ are related by

$$q(u) = -p(u) - up'(u) \tag{17}$$

(see Barndorff-Nielsen 1998).

The logarithm of the characteristic function of a random vector $(X(t_1), ..., X(t_m))$ is of the form

$$\log E \exp\{i(z_1 X(t_1) + \ldots + z_m X(t_m)\}$$

$$= \int_{\mathbb{R}} \kappa(\sum_{j=1}^{m} z_j e^{-\lambda(t_j - s)} \mathbf{1}_{[0,\infty)}(t_j - s)) ds, \tag{18}$$

where

$$\kappa(z) = \log E \exp\{i z Z(1)\} = C\{z; Z(1)\},$$

and the function (18) has the form (12) with Lévy triplet $(\tilde{a}, \tilde{d}, \tilde{v})$ of $Z(1)$.

The logarithms of the moment generation functions (if they exist) take the forms

$$\log M(\zeta) = \log E \exp\{\zeta X(t)\} = \zeta a + \frac{d}{2}\zeta^2 + \int_{\mathbb{R}}(e^{\zeta u} - 1 - \zeta u \mathbf{1}_{[-1,1]}(u))v(du),$$

where the triplet $(a, d, v)$ is the Lévy triplet of $X(0)$, or in terms of the Lévy triplet $(\tilde{a}, \tilde{d}, \tilde{v})$ of $Z(1)$

$$\log M(\zeta) = \tilde{a} \int_{\mathbb{R}} (\zeta e^{-\lambda(t-s)} \mathbf{1}_{[0,\infty)}(t-s)) ds + \frac{\tilde{d}}{2}\zeta^2 \int_{\mathbb{R}} (\zeta e^{-\lambda(t-s)} \mathbf{1}_{[0,\infty)}(t-s))^2 ds$$

$$+ \int_{\mathbb{R}} \int_{\mathbb{R}} \left[ \exp\left\{ u\zeta e^{-\lambda(t-s)} \mathbf{1}_{[0,\infty)}(t-s) \right\} \right.$$
$$\left. -1 - u\left( \zeta e^{-\lambda(t-s)} \mathbf{1}_{[0,\infty)}(t-s) \right) \mathbf{1}_{[-1,1]}(u) \right] \tilde{v}(du) ds, \tag{19}$$

and

$$\log M(\zeta_1, \zeta_2; t_1 - t_2) = \log E \exp\{\zeta_1 X(t_1) + \zeta_2 X(t_2)\}$$

$$= \tilde{a} \int_{\mathbb{R}} (\sum_{j=1}^{2} \zeta_j e^{-\lambda(t_j - s)} \mathbf{1}_{[0,\infty)}(t_j - s)) ds$$

$$+ \frac{\tilde{d}}{2}\zeta^2 \int_{\mathbb{R}} (\sum_{j=1}^{2} \zeta_j e^{-\lambda(t_j - s)} \mathbf{1}_{[0,\infty)}(t_j - s))^2 ds$$

$$+ \int_{\mathbb{R}} \int_{\mathbb{R}} \left[ \exp\left\{ u \sum_{j=1}^{2} \zeta_j e^{-\lambda(t_j - s)} \mathbf{1}_{[0,\infty)}(t_j - s) \right\} - 1 \right.$$
$$\left. -u\left( \sum_{j=1}^{2} \zeta_j e^{-\lambda(t_j - s)} \mathbf{1}_{[0,\infty)}(t_j - s) \right) \mathbf{1}_{[-1,1]}(u) \right] \tilde{v}(du) ds. \tag{20}$$

The following result is needed in our construction of multifractal Ornstein-Uhlenbeck processes.

**Theorem 3.1.** *Let $X(t), t \in \mathbb{R}_+$ be an OU-type stationary process (14) such that the Lévy measure $v$ in (12) of the random variable $X(t)$ satisfies the condition that for some $q \in Q \subseteq \mathbb{R}$,*

$$\int_{|x| \geq 1} g(x)v(dx) < \infty, \tag{21}$$

where $g(x)$ denotes any of the functions $e^{2qx}, e^{qx}, e^{qx}|x|$. Then, for the geometric OU-type process $\Lambda_q(t) := e^{qX(t)}$,

$$\sum_{n=0}^{\infty} c(q, b^{-n}) < \infty,$$

where $c(q,t) = E\sup_{s\in[0,t]} |\Lambda_q(0)^q - \Lambda_q(s)^q|$.

The proof of Theorem 2 is given in Anh, Leonenko and Shieh (2008). To prove that a geometric OU-type process satisfies the covariance decay condition of equation (6) in Theorem 1, the expression given by (20) is not ready to yield the decay as $t_2 - t_1 \to \infty$. The following proposition gives a general decay estimate which the driving Lévy processes $Z$ in Section 4 below indeed satisfy.

**Proposition 0.1.** *Consider the stationary OU-type process $X$ defined by*

$$dX(t) = -\lambda X(t)dt + dZ(\lambda t),$$

*which has a stationary distribution $\pi(x)$ such that, for some $a > 0$,*

$$\int |x|^a \pi(dx) < \infty. \tag{22}$$

*Then there exist positive constants $c$ and $C$ such that*

$$\mathrm{Cov}\left(e^{X(t)}, e^{X(0)}\right) \leq Ce^{-ct}$$

*for all $t > 0$.*

*Proof.* Masuda (2004) showed that, under the assumption (22), the stationary process $X(t)$ satisfies the $\beta$-mixing condition with coefficient $\beta_X(t) = O(e^{-ct}), t > 0$. Note that this is also true for the stationary process $e^{X(t)}$, since the $\sigma$-algebras generated by these two processes are equivalent. Hence,

$$\beta_{e^X}(t) = O(e^{-ct}), \quad t > 0.$$

It then follows that

$$\mathrm{Cov}\left(e^{X(t)}, e^{X(0)}\right) \leq const \times \beta_{e^X}(t) \leq Ce^{-ct}$$

(see Billingsley 1968).

The above Proposition is a corrected version of Proposition 1 in Anh, Leonenko and Shieh (2008). We note that Jongbloed *et al.* (2005) proved that the process $X$ also satisfies the $\beta$-mixing condition under the condition

$$\int_2^{\infty} \log(x)\nu(dx) < \infty, \tag{23}$$

where $v$ is the Lévy measure of the OU process $X(t)$. Condition (22) is stronger than condition (23).

The correlation structures found in applications may be more complex than the exponential decreasing autocorrelation of the form (16). Barndorff-Nielsen (1998), Barndorff-Nielsen and Sheppard (2001), Barndorff-Nielsen and Leonenko (2005) proposed to consider the following class of autocovariance functions:

$$R_{\text{sup}}(t) = \sum_{j=1}^{m} \sigma_j^2 \exp\{-\lambda_j |t|\}, \tag{24}$$

which is flexible and can be fitted to many autocovariance functions arising in applications.

In order to obtain models with dependence structure (24) and given marginal density with finite variance, we consider stochastic processes defined by

$$dX_j(t) = -\lambda_j X_j(t)\,dt + dZ_j(\lambda_j t), \; j = 1, 2, ..., m,$$

and their superposition

$$X_{\text{sup}}(t) = X_1(t) + ... + X_m(t), \; t \geq 0, \tag{25}$$

where $Z_j$, $j = 1, 2, ..., m$, are mutually independent Lévy processes. Then the solution $X_j = \{X_j(t), t \geq 0\}$, $j = 1, 2, ..., m$, is a stationary process. Its correlation function is of the exponential form (assuming finite variance).

The superposition (25) has its marginal density given by that of the random variable

$$X_{\text{sup}}(0) = X_1(0) + ... + X_m(0), \tag{26}$$

autocovariance function (24) (where $\sigma_j^2$ are now variances of $X_j$), and spectral density

$$f_{\text{sup}}(\lambda) = \frac{2}{\pi} \sum_{j=1}^{m} \sigma_j^2 \frac{\lambda_j}{\lambda_j + \lambda^2}, \; \lambda \in \mathbb{R}.$$

We are interested in the case when the distribution of (26) is tractable, for instance when $X_{\text{sup}}(0)$ belongs to the same class as $X_j(0), j = 1, ..., m$ (see the example in Subsection 4.1 below).

Note that an infinite superposition ($m \to \infty$) gives a complete monotone class of covariance functions

$$R_{\text{sup}}(t) = \int_0^\infty e^{-tu} dU(u), t \geq 0,$$

for some finite measure $U$, which display long-range dependence (see Barndorff-Nielsen 1998, 2001, Barndorff-Nielsen and Leonenko 2005 for possible covariance structures and spectral densities).

# 4 Multifractal Ornstein-Uhlenbeck processes

This section introduces two new scenarios which generalize the log-inverse Gaussian scenario and the log-normal inverse Gaussian scenario obtained in Anh, Leonenko and Shieh (2008).

## 4.1 Log-tempered stable scenario

This subsection constructs a multifractal process based on the geometric tempered stable OU process. In this case, the mother process takes the form $\Lambda(t) = \exp\{X(t) - c_X\}$, where $X(t)$ is a stationary OU type process (15) with tempered stable distribution and $c_X$ is a constant depending on the parameters of its marginal distribution. This form is needed for the condition $E\Lambda(t) = 1$ to hold. The log-tempered stable scenario appeared in Novikov (1994) and Anh *et al.* (2001) in a physical setting and under different terminology. We provide a rigorous construction for this scenario here.

We consider the stationary OU process whose marginal distribution is the tempered stable distribution $TS(\kappa, \delta, \gamma)$ (see, for example, Barndorff-Nielsen and Shephard 2002). This distribution is the exponentially tilted version of the positive $\kappa-$stable law $S(\kappa, \delta)$ whose probability density function (pdf) is

$$s_{\kappa,\delta}(x) = \frac{\delta^{-1/\kappa}}{2\pi} \sum_{k=1}^{\infty} (-1)^{k-1} \sin(k\pi\kappa) \frac{\Gamma(k\kappa+1)}{k!} 2^{k\kappa+1} \left(\frac{x}{\delta^{1/\kappa}}\right)^{-k\kappa-1}, \quad (27)$$

$$x > 0, \kappa \in (0,1), \delta > 0.$$

The pdf of the tempered stable distribution $TS(\kappa, \delta, \gamma)$ is

$$\pi(x) = \pi(x; \kappa, \delta, \gamma) = e^{\delta\gamma} s_{\kappa,\delta}(x) e^{-\frac{x}{2}\gamma^{1/\kappa}}, \quad x > 0, \quad (28)$$

where the parameters satisfy

$$\kappa \in (0,1), \delta > 0, \gamma > 0.$$

It is clear that $TS(\frac{1}{2}, \delta, \gamma) = IG(\delta, \gamma)$, the inverse Gaussian distribution with pdf

$$\pi(x) = \frac{1}{\sqrt{2\pi}} \frac{\delta e^{\delta\gamma}}{x^{3/2}} \exp\left\{-\left(\frac{\delta^2}{x} + \gamma^2 x\right)\frac{1}{2}\right\} \mathbf{1}_{[0,\infty)}(x), \delta > 0, \gamma \geq 0,$$

and $TS(\frac{1}{3}, \delta, \gamma)$ has the pdf

$$\pi(x) = \frac{\sqrt{2}}{\pi} \delta^{3/2} e^{\delta^{3/2\gamma}} x^{-3/2} K_{\frac{1}{3}}\left(\left(\frac{2\delta}{3}\right)^{3/2} \frac{1}{\sqrt{x}}\right) e^{-\frac{x\gamma^3}{2}}, x > 0,$$

where here and below

$$K_\lambda(z) = \int_0^\infty e^{-z\cosh(u)} \cosh(uz)du, \, z > 0, \tag{29}$$

is the modified Bessel function of the third kind of index $\lambda$.

The cumulant transform of a random variable $X$ with pdf (28) is equal to

$$\log \mathrm{E}e^{\zeta X} = \delta\gamma - \delta\left(\gamma^{\frac{1}{\kappa}} - 2\zeta\right)^\kappa, 0 < \zeta < \frac{\gamma^{1/\kappa}}{2}. \tag{30}$$

We will consider a stationary OU type process (15) with marginal distribution $TS(\kappa, \delta, \gamma)$. This distribution is self-decomposable (and hence infinitely divisible) with the Lévy triplet $(a, 0, v)$, where

$$v(du) = b(u)du, \tag{31}$$

$$b(u) = 2^\kappa \delta \frac{\kappa}{\Gamma(1-\kappa)} u^{-1-\kappa} e^{-\frac{u\gamma^{1/\kappa}}{2}}, u > 0.$$

Thus, if $X_{j,}(t)$, $j = 1, ..., m$, are independent so that $X_j(t) \sim TS(\kappa, \delta_j, \gamma)$, $j = 1, ..., m$, then we have that

$$X_1(t) + ... + X_m(t) \sim TS(\kappa, \sum_{j=1}^m \delta_j, \gamma).$$

The BDLP $Z(t)$ in (15) has a Lévy triplet $(\tilde{a}, 0, \tilde{v})$, with

$$\tilde{v}(du) = \lambda\omega(u)du,$$

$$\omega(u) = 2^\kappa \delta \frac{\kappa}{\Gamma(1-\kappa)} \left(\frac{\kappa}{u} + \frac{\gamma^{1/\kappa}}{2}\right) u^{-\kappa} e^{-\frac{u\gamma^{1/\kappa}}{2}}, u > 0, \tag{32}$$

that is, the BDLP of the distribution $TS(\kappa, \delta, \gamma)$ is the sum of a Lévy process with density

$$2^\kappa \delta \frac{\kappa^2}{\Gamma(1-\kappa)} u^{-1-\kappa} e^{-\frac{u\gamma^{1/\kappa}}{2}}, u > 0,$$

and a compound Poisson process $Z(t) = \sum_{k=1}^{P(t)} Z_k$, which has Lévy density

$$\delta\gamma\kappa \frac{(\gamma^{\frac{1}{\kappa}}/2)^{1-\kappa}}{\Gamma(1-\kappa)} u^{-\kappa} e^{-\frac{u\gamma^{1/\kappa}}{2}}, u > 0,$$

meaning that the summands $Z_k$ have gamma distribution $\Gamma(1-\kappa, \frac{\gamma^{1/\kappa}}{2})$, while the independent Poisson process has rate $\delta\gamma\kappa$.

The correlation function of the stationary process $X(t)$ then takes the form

$$r_X(t) = \exp\{-\lambda|t|\}, \ t \in \mathbb{R},$$

and

$$EX(t) = 2\kappa\delta\gamma^{\frac{\kappa-1}{\kappa}}, \operatorname{Var}X(t) = 4\kappa(1-\kappa)\delta\gamma^{\frac{\kappa-2}{\kappa}}.$$

**B'**. Consider a mother process of the form

$$\Lambda(t) = \exp\{X(t) - c_X\},$$

with

$$c_X = \delta\gamma - \delta\left(\gamma^{\frac{1}{\kappa}} - 2\right)^{\kappa}, \gamma > 2^{\kappa},$$

where $X(t), t \in \mathbb{R}_+$ is a stationary $TS(\kappa, \delta, \gamma)$ OU-type process with covariance function

$$R_X(t) = \operatorname{Var}X(t)\exp\{-\lambda|t|\}, t \in \mathbb{R}.$$

Under condition **B'**, we obtain the following moment generating function:

$$M(\zeta) = E\exp\{\zeta(X(t) - c_X)\} = e^{-c_X\zeta}e^{\delta\gamma - \delta\left(\gamma^{\frac{1}{\kappa}} - 2\zeta\right)^{\kappa}}, \quad 0 < \zeta < \frac{\gamma^{1/\kappa}}{2}, \gamma > 2^{\kappa}, \tag{33}$$

and bivariate moment generating function:

$$M(\zeta_1, \zeta_2; (t_1 - t_2)) = E\exp\{\zeta_1(X(t_1) - c_X) + \zeta_2(X(t_2) - c_X)\}$$

$$= e^{-c_X(\zeta_1+\zeta_2)}E\exp\{\zeta_1 X(t_1) + \zeta_2(X(t_2))\}, \tag{34}$$

where $\theta = (\kappa, \delta, \gamma)$, and $E\exp\{\zeta_1 X(t_1) + \zeta_2(X(t_2))\}$ is given by (19) with Lévy measure $\tilde{\nu}$ having density (32). Thus, the correlation function of the mother process takes the form

$$\rho(\tau) = \frac{M(1, 1; \tau) - 1}{M(2) - 1}. \tag{35}$$

We can formulate the following

**Theorem 4.1.** *Suppose that condition* **B'** *holds,* $\lambda > 0$ *and* $q \in Q = \{q : 0 < q < \frac{\gamma^{1/\kappa}}{2}, \gamma > 4^{\kappa}\}$. *Then, for any*

$$b > \exp\left(2\delta\left(\gamma^{\frac{1}{\kappa}} - 2\right)^{\kappa} - \delta\left(\gamma^{\frac{1}{\kappa}} - 4\right)^{\kappa} - \delta\gamma\right), \gamma > 4^{\kappa},$$

the stochastic processes defined by (2) converge in $L_2$ to the stochastic process $A(t)$ for each fixed $t$ as $n \to \infty$ such that, if $A(1) \in L_q$ for $q \in Q$, then

$$EA^q(t) \sim t^{T(q)+1},$$

where the Rényi function $T(q)$ is given by

$$T(q) = \left(1 + \frac{\delta\gamma}{\log b} - \frac{\delta}{\log b}\left(\gamma^{\frac{1}{\kappa}} - 2\right)^{\kappa}\right)q + \frac{\delta}{\log b}\left(\gamma^{\frac{1}{\kappa}} - 2q\right)^{\kappa} - \frac{\delta\gamma}{\log b} - 1.$$

Moreover,

$$VarA(t) \geqslant \int_0^t \int_0^t [M(1,1;u-w) - 1]\,dudw,$$

where $M$ is given by (34).

Theorem 3 follows from Theorems 1&2 and Proposition 1. Note that for $\kappa = 1/2$ Theorem 3 is identical to Theorem 4 of Anh, Leonenko and Shieh (2008).

We can construct log-tempered stable scenarios for a more general class of finite superpositions of stationary tempered stable OU-type processes:

$$X_{\sup}(t) = \sum_{j=1}^m X_j(t), t \in \mathbb{R}_+,$$

where $X_j(t), j = 1,...,m$, are independent stationary processes with marginals $X_j(t) \sim TS(\kappa, \delta_j, \gamma), j = 1,...,m$, and parameters $\delta_j, j = 1,...,m$. Then $X_{\sup}(t)$ has the marginal distribution $TS(\kappa, \sum_{j=1}^m \delta_j, \gamma)$, and covariance function

$$R_{\sup}(t) = 4\kappa(1-\kappa)\gamma^{\frac{\kappa-2}{\kappa}}\sum_{j=1}^m \delta_j \exp\left\{-\lambda_j |t|\right\}, t \in \mathbb{R}.$$

Theorem 2 and Proposition 1 can be generalized to this situation and the statement of Theorem 3 can be reformulated for $X_{\sup}$ with $\delta = \sum_{j=1}^m \delta_j$ , and $M(\zeta_1, \zeta_2; (t_1 - t_2)) = \prod_{j=1}^m M_j(\zeta_1, \zeta_2; (t_1 - t_2))$, $\theta_j = (\kappa, \delta_j, \gamma)$, and $\lambda$ being replaced by $\lambda_j$ in the expression (20) for $M_j(\zeta_1, \zeta_2; (t_1 - t_2))$.

### 4.2 Log-normal tempered stable scenario

This subsection constructs a multifractal process based on the geometric normal tempered stable (NTS) OU process. We consider a random variable $X = \mu + \beta Y + \sqrt{Y}\varepsilon$, where the random variable $Y$ follows the $TS(\kappa, \delta, \gamma)$ distribution, $\varepsilon$ has a standard normal distribution, and $Y$ and $\varepsilon$ are independent. We then say that $X$ follows the normal tempered stable law $NTS(\kappa, \gamma, \beta, \mu, \delta)$ (see, for example, Barndorff-Nielsen and Shephard 2002). In particular, for $\kappa = 1/2$ we have that $NTS(\frac{1}{2}, \gamma, \beta, \mu, \delta)$ is the same as the normal inverse Gaussian law $NIG(\alpha, \beta, \mu, \delta)$ with $\alpha = \sqrt{\beta^2 + \gamma^2}$ (see Barndorff-Nielsen 1998). Assuming $\mu = 0$ for simplicity, the probability density function of $NTS(\kappa, \gamma, \beta, \mu, \delta)$ may be written as a mixture representation

$$p(x; \kappa, \gamma, \beta, 0, \delta) = \frac{1}{\sqrt{2\pi}} e^{\delta\gamma + \beta x} \int_0^\infty z^{-1/2} e^{-\frac{1}{2}(\frac{x^2}{z} + \beta^2 z)} \pi(z; \kappa, \delta, \gamma) dz, x \in \mathbb{R}, \quad (36)$$

where $\pi(z; \kappa, \delta, \gamma)$ is the pdf of the $TS(\kappa, \delta, \gamma)$ distribution. We assume that

$$\mu \in \mathbb{R}, \delta > 0, \gamma > 0, \beta > 0, \kappa \in (0, 1).$$

Let $\alpha = \sqrt{\beta^2 + \gamma^{1/\kappa}}$ and using the substitution $s = 1/z$ we obtain

$$p(x; \kappa, \gamma, \beta, 0, \delta) = \frac{1}{\sqrt{2\pi}} e^{\delta\gamma + \beta x} \int_0^\infty s^{-3/2} e^{-\frac{1}{2}(x^2 s + \alpha^2/s)} s_{\kappa,\delta}(1/s) ds, \ x \in \mathbb{R},$$

where $s_{\kappa,\delta}(x)$ is given by (27). As an example, the pdf of $NTS(\frac{1}{3}, \alpha, \beta, 0, 1)$ has the form

$$p(x; \frac{1}{3}, \alpha, \beta, 0, 1) = (\frac{\delta}{\pi})^{\frac{3}{2}} e^{\delta\gamma + \beta x} \int_0^\infty e^{-\frac{1}{2}(x^2 s + \alpha^2/s)} K_{\frac{1}{3}}\left(\left(\frac{2\delta}{3}\right)^{3/2} \sqrt{s}\right) ds.$$

It was pointed out by Barndorff-Nielsen and Shephard (2002) that $NTS(\kappa, \gamma, \beta, \mu, \delta)$ is self-decomposable.

Also one can obtain the following semiheavy tail behaviour:

$$p(x; \kappa, \gamma, \beta, 0, \delta) \sim 2^{\kappa+1} \delta e^{\delta\gamma} \frac{\Gamma(1+\kappa)}{\Gamma(\kappa)\Gamma(1-\kappa)} \alpha^{\kappa+\frac{1}{2}} |x|^{-\kappa-1} e^{-\alpha|x|+\beta x}$$

as $x \to \pm\infty$.

The cumulant transform of the random variable $X$ with pdf (36) is equal to

$$\log E e^{\zeta X} = \mu\zeta + \delta\gamma - \delta\left(\alpha^2 - (\beta + \zeta)^2\right)^\kappa, |\beta + \zeta| < \alpha = \sqrt{\beta^2 + \gamma^{1/\kappa}}. \quad (37)$$

The Lévy triplet of $NTS(\kappa, \gamma, \beta, \mu, \delta)$ is $(a, 0, v)$, where

$$v(du) = b(u)du, \quad (38)$$

$$b(u) = \frac{\delta}{\sqrt{2\pi}} \alpha^{\kappa+\frac{1}{2}} \frac{\kappa 2^{\kappa+1}}{\Gamma(1-\kappa)} |u|^{-(\kappa+\frac{1}{2})} K_{\kappa+\frac{1}{2}}(\alpha|u|) e^{\beta u}, u \in \mathbb{R},$$

the modified Bessel function $K_\lambda(z)$ being given by (29). Thus, if $X_j(t)$, $j = 1, ..., m$, are independent so that $X_j(t) \sim NTS(\kappa, \gamma, \beta, \mu_j, \delta_j)$, $j = 1, ..., m$, then we have that

$$X_1(t) + ... + X_m(t) \sim NTS(\kappa, \gamma, \beta, \sum_{j=1}^{m} \mu_j, \sum_{j=1}^{m} \delta_j).$$

From (17), (37) and the formulae

$$K_\lambda(x) = K_\lambda(-x), \ K_{-\lambda}(x) = K_\lambda(x),$$

$$\frac{d}{dx} K_\lambda(x) = -\frac{\lambda}{x} K_\lambda(x) - K_{\lambda-1}(x),$$

we obtain that the BDLP $Z(t)$ in (15) has a Lévy triplet $(\widetilde{a}, 0, \widetilde{v})$, with

$$\widetilde{v}(du) = \lambda \omega(u) du,$$

where

$$\omega(u) = -b(u) - ub'(u) = \frac{\delta}{\sqrt{2\pi}} \alpha^{\kappa+\frac{1}{2}} \frac{\kappa 2^{\kappa+1}}{\Gamma(1-\kappa)}$$

$$\times \{(\kappa - \frac{1}{2}) |u|^{-(\kappa+\frac{1}{2})} K_{\kappa+\frac{1}{2}}(\alpha |u|) e^{\beta u} + |u|^{-(\kappa-\frac{1}{2})} [-\frac{\kappa+\frac{1}{2}}{|u|} K_{\kappa+\frac{1}{2}}(\alpha |u|) e^{\beta u}$$

$$-K_{\kappa-\frac{1}{2}}(\alpha |u|) e^{\beta u} \alpha + K_{\kappa+\frac{1}{2}}(\alpha |u|) e^{\beta u} \beta]\}. \tag{39}$$

The correlation function of the stationary process $X(t)$ then takes the form

$$r_X(t) = \exp\{-\lambda |t|\}, t \in \mathbb{R},$$

and

$$EX(t) = \exp\left(\gamma\delta - \delta(\alpha^2 - \beta^2)^\kappa\right) \left(\mu + 2\kappa\beta\delta(\alpha^2 - \beta^2)^{\kappa-1}\right),$$

$$VarX(t) = \exp\left(\gamma\delta - \delta(\alpha^2 - \beta^2)^\kappa\right)$$

$$\times \left\{2\kappa\delta(\alpha^2 - \beta^2)^{\kappa-1} - 4\kappa\beta^2\delta(\kappa-1)(\alpha^2 - \beta^2)^{\kappa-2}\right.$$

$$\left. + \left(\mu + 2\kappa\beta\delta(\alpha^2 - \beta^2)^{\kappa-1}\right)^2\right\}$$

$$-\exp\left(2\gamma\delta - 2\delta(\alpha^2 - \beta^2)^\kappa\right) \left(\mu + 2\kappa\beta\delta(\alpha^2 - \beta^2)^{\kappa-1}\right)^2.$$

We see that the variance cannot be factorized in a similar manner as in Subsection 4.1 and thus superposition cannot be used to create multifractal scenarios with more elaborate dependence structures.

**B''.** Consider a mother process of the form

$$\Lambda(t) = \exp\{X(t) - c_X\},$$

with

$$c_X = \mu + \delta\gamma - \delta\left(\alpha^2 - (\beta+1)^2\right)^\kappa, |\beta+1| < |\alpha|,$$

where $X(t), t \in \mathbb{R}_+$ is a stationary $NTS(\kappa, \gamma, \beta, \mu, \delta)$ OU-type process with covariance function

$$R_X(t) = \operatorname{Var}X(t)\exp\{-\lambda|t|\}, t \in \mathbb{R}.$$

Under condition **B''**, we obtain the following moment generating function

$$M(\zeta) = \operatorname{E}\exp\{\zeta(X(t) - c_X)\} = e^{-c_X\zeta}e^{\mu\zeta + \delta\gamma - \delta(\alpha^2 - (\beta+\zeta)^2)^\kappa}, \quad |\beta+\zeta| < \alpha, \tag{40}$$

and bivariate moment generating function

$$M(\zeta_1, \zeta_2; (t_1 - t_2)) = \operatorname{E}\exp\{\zeta_1(X(t_1) - c_X) + \zeta_2(X(t_2) - c_X)\}$$

$$= e^{-c_X(\zeta_1+\zeta_2)}\operatorname{E}\exp\{\zeta_1 X(t_1) + \zeta_2(X(t_2))\}, \tag{41}$$

where $\theta = (\kappa, \delta, \gamma)$, and $\operatorname{E}\exp\{\zeta_1 X(t_1) + \zeta_2(X(t_2))\}$ is given by (19) with Lévy measure $\tilde{\nu}$ having density (39). Thus, the correlation function of the mother process takes the form

$$\rho(\tau) = \frac{M(1, 1; \tau) - 1}{M(2) - 1}. \tag{42}$$

We can formulate the following

**Theorem 4.2.** *Suppose that condition **B''** holds, $\lambda > 0$ and*

$$q \in Q = \left\{q; |q + \beta| < \alpha = \sqrt{\beta^2 + \gamma^{1/\kappa}}, |\beta+1| < \alpha, |\beta+2| < \alpha\right\}.$$

*Then, for any*

$$b > \exp\left\{\delta[(2\alpha^2 - (\beta+1)^2)^\kappa - (\alpha^2 - (\beta+2)^2)^\kappa - \gamma]\right\},$$

*the stochastic processes defined by (2) converge in $L_2$ to the stochastic process $A(t)$ for each fixed $t$ as $n \to \infty$ such that, if $A(1) \in L_q$ for $q \in Q$, then*

$$\operatorname{E}A^q(t) \sim t^{T(q)+1},$$

*where the Rényi function $T(q)$ is given by*

$$T(q) = \left(1 + \frac{c_X}{\log b} - \frac{\mu}{\log b}\right) q + \frac{\delta}{\log b}\left(\alpha^2 - (\beta + q)^2\right)^\kappa - \frac{\delta\gamma}{\log b} - 1.$$

*Moreover,*

$$VarA(t) \geqslant \int_0^t \int_0^t [M(1,1; u - w) - 1]\, du\, dw,$$

*where M is given by (41).*

Theorem 4 follows from Theorems 1&2 and Proposition 1, and the following formula for $z \to \infty$ :

$$K_\lambda(z) = \sqrt{\frac{\pi}{2}} z^{-1/2} e^{-z}\left(1 + \frac{4\lambda^2 - 1}{8z} + \ldots\right), \ z > 0.$$

Note that, for $\kappa = 1/2$, Theorem 4 is identical to Theorem 5 of Anh, Leonenko and Shieh (2008).

**Acknowledgements** Partially supported by the Australian Research Council grant DP0559807, the National Science Foundation grant DMS-0417676, the EPSRC grant EP/D057361 (RCMT 119), the Marie Curie grant of European Communities PIRSES-GA-2008-230804 (RCMT 152), and the Taiwan NSC grant 962115M002005MY3. Leonenko's research was partially supported by the Welsh Institute of Mathematics and Computational Ssciences. The authors wish to thank the Editor, the referees and Dr Ross McVinish for their constructive comments and suggestions.

# References

[1] P. Abry, R. Baraniuk, P. Flandrin, R. Riedi and D. Veitch, The multiscale nature of network traffic: Discovery, analysis, and modelling, *IEEE Signal Processing Magazine* 19(2002), 28-46.

[2] V. V. Anh, K.-S. Lau and Z.-G. Yu, Multifractal characterization of complete genomes, *Journal of Physics A: Mathematical and General* 34(2001), 7127–7139.

[3] V. V. Anh and N. N. Leonenko, Non-Gaussian scenarios for the heat equation with singular initial data, *Stochastic Processes and their Applications* 84(1999), 91–114.

[4] V. V. Anh, N. N. Leonenko and N.-R. Shieh, Multifractality of products of geometric Ornstein-Uhlenbeck type processes, *Advances of Applied Probablity* 40(4) (2008), 1129-1156.

[5] V.V. Anh, N.N. Leonenko and N.-R. Shieh, Multifractal scaling of products of birth-death processes, *Bernoulli* (2009a), DOI 10.3150/08-BEJ156.

[6] V.V. Anh, N.N. Leonenko and N.-R. Shieh, Multifractal products of stationary diffusion processes, *Stochastic Analysis and Applications* 27 (2009b), DOI:10.1080/07362990802679091.

[7] D. Applebaum, Lévy Processes and Stochastic Calculus, Cambridge University Press, Cambridge, 2004.

[8] E. Bacry and J.F. Muzy, Log-infinitely divisible multifractal processes, *Comm. Math. Phys.* 236 (2003), 449–475.

[9] O. E. Barndorff-Nielsen, Processes of normal inverse Gaussian type, *Finance and Stochastics* 2(1998), 41–68.

[10] O. E. Barndorff-Nielsen, Superpositions of Ornstein-Uhlenbeck type processes, *Theory Probab. Applic.* 45(2001), 175–194.

[11] O.E. Barndorff-Nielsen and N.N. Leonenko, Spectral properties of superpositions of Ornstein-Uhlenbeck type processes, *Methodol. Comput. Appl. Probab.* 7 (2005), 335–352.

[12] O. E. Barndorff-Nielsen and N. Shephard, Non-Gaussian Ornstein-Uhlenbeck-based models and some of their uses in financial economics, part 2, *J. R. Stat. Soc. B* 63 (2001), 167–241.

[13] O.E. Barndorff-Nielsen and N. Shephard, Normal modified stable processes, *Theory Probab. Math. Statist.* 65 (2002), 1–20.

[14] J. Barral and B. Mandelbrot, Multiplicative products of cylindrical pulses, *Prob. Th. Rel. Fields* 124(2002), 409–430.

[15] J. Bertoin, Lévy Processes, Cambridge University Press, 1996.

[16] P. Chainais, R. Riedi and P. Abry, On non-scale-invariant infinitely divisible cascades, *IEEE Trans. Infor. Theory* 51 (2005), 1063–1083.

[17] C. Donati-Martin and M.Yor, Some exlicit Krein representations of certain subordinators, including the Gamma process, Pbl. RIMS, Kyoto University, 42(2006), 879-895.

[18] K. Falconer, Techniques in Fractal Geometry, Wiley, 1997.

[19] U. Frisch, Turbulence, Cambridge University Press, 1995.

[20] V. K. Gupta and E. C. Waymire, A statistical analysis of mesoscale rainfall as a random cascade, *Journal of Applied Meteorology* 32(1993), 251–267.

[21] N. Hohn, D. Veitch and P. Abry, Cluster processes: a natural language for network traffic, *IEEE Trans. Signal Process.* 51 (2003), 2229–2244.

[22] S. Jaffard, The multifractal nature of Lévy processes, *Probab. Theory Relat. Fields* 114(1999), 207–227.

[23] G. Jongbloed, F.H. van der Meulen and A.W. van der Vaart, Nonparametric inference for Levy-driven OU process, *Bernoulli*, 11 (2005), 759–791.

[24] J.-P. Kahane, Sur la chaos multiplicatif, *Ann. Sc. Math. Québec* 9(1985), 105–150.

[25] J.-P. Kahane, Positive martingale and random measures, *Chinese Annals of Mathematics* 8B(1987), 1–12.

[26] A. N. Kolmogorov, Local structure of turbulence in fluid for very large Reynolds numbers, *Doklady Acad. of Sciences of USSR* 31(1941), 538–540.

[27] A. N. Kolmogorov, On refinement of previous hypotheses concerning the local structure in viscous incompressible fluid at high Reynolds number, *J. Fluid Mech.* 13(1962), 82–85.

[28] A. E. Kyprianou, Introductory Lectures on Fluctuations of Lévy Processes with Applications, Universitext, Springer-Verlag, Berlin, 2006.

[29] B. Mandelbrot, Fractals and Scaling in Finance, Springer, New York, 1997.

[30] P. Mannersalo, I. Norros and R. Riedi, Multifractal products of stochastic processes: construction and some basic properties, *Adv. Appl. Prob.* 34(2002), 888–903.

[31] P. Mörters and N.-R. Shieh, Thin and thick points for branching measure on a Galton-Watson tree, *Statist. Prob. Letters* 58(2002), 13-22.

[32] P. Mörters and N.-R. Shieh, On multifractal spectrum of the branching measure on a Galton-Watson tree, *J. Appl. Prob.* 41(2004), 1223-1229.

[33] P. Mörters and N.-R. Shieh, Multifractal analysis of branching measure on a Galton-Watson tree, Proceedings of the 3rd International Congress of Chinese Mathematicians, pp 655-662, ed. K.S. Lau, Z.P. Xin, and S.T. Yau. AMS/IP Studies in Advanced Math **42**, 2008.

[34] E. A. Novikov, Infinitely divisible distributions in turbulence, *Physical Review E* 50(1994), R3303-R3305.

[35] R. Riedi, Multifractal processes, In P. Doukhan, G. Oppenheim and M. Taqqu, editors, Theory and Applications of Long-Range Dependence, pp. 625–716, Birkhäuser, Boston, 2003.

[36] K. Sato, Lévy Processes and Infinitely Divisible Distributions, Cambridge University Press, 1999.

[37] D. Schertzer and S. Lovejoy, Space-time complexity and multifractal predictability, *Phys. A* 338(2004), 173–186.

[38] D. Schertzer, S. Lovejoy, F. Schmitt, Y. Chigirinskaya and D. Marsan, Multifractal cascade dynamics and turbulent intermittency, *Fractals* 5(1997), 427–471.

[39] F.G. Schmitt, A causal multifractal stochastic equation and its statistical properties, *European Physical Journal* B 34(2003), 85-98.

[40] F.G. Schmitt and D. Marsan, Stochastic equation generating continuous multiplicative cascades, *Eur. J. Phys.* B 20(2001), 3-6.

[41] N.-R. Shieh and S. J. Taylor, Multifractal spectra of branching measure on a Galton-Watson tree, *J. Appl. Prob.* 39(2002), 100-111.

[42] A.V. Skorokhod, Random Processes with Independent Increments, Kluwer, Dordrecht, 1991.

[43] G.Terdik and W.A.Woyczynski (2004) Rosinski measures for tempered stable and related Ornstein-Uhlenbeck processes, Probabality and Mathematical Statistics, 26 (2004), 213-243.

[44] J.Y. Yang, Z.G. Yu and V.V. Anh, Clustering structure of large proteins using multifractal analyses based on 6-letter model and hydrophobicity scale of amino acids, *Chaos, Solitons and Fractals*, doi:10.1016/j.chao.2007.08.014.

# A new look at measuring dependence

Wei Biao Wu and Jan Mielniczuk

**Abstract** This paper revisits the concept of dependence. We view statistical dependence as the state of variables being influenced by others. Our viewpoint accords well with the daily understanding of the notion of dependence, while classical dependence measures such as Pearson's correlation coefficient, Kendall's $\tau$ and Spearman's $\rho$ have different meanings. With this understanding of dependence, we introduce new dependence measures which are easy to work with and they are useful for developing an asymptotic theory for complicated stochastic systems. We also explore relations of the introduced dependence concept with nonlinear system theory, experimental design, information theory and risk management.

## 1 Introduction

According to the Merriam-Webster dictionary, *dependence* means that

"the quality or state of being dependent; especially: the quality or state of being influenced or determined by or subject to another."

This interpretation appears to accord well with our daily understanding of the notion of dependence. In probability theory and statistics, however, the meaning of dependence is understood somewhat differently. As pointed out in Drouet Mari and Kotz (2001, p. 31):

Wei Biao Wu
University of Chicago,
Department of Statistics, The University of Chicago,
Chicago, IL 60637, USA, e-mail: wbwu@galton.uchicago.edu

Jan Mielniczuk
Polish Academy of Sciences
Warsaw University of Technology, Warsaw, Poland
e-mail: miel@ipipan.waw.pl

P. Doukhan et al. (eds.), *Dependence in Probability and Statistics*,
Lecture Notes in Statistics 200, DOI 10.1007/978-3-642-14104-1_7,
© Springer-Verlag Berlin Heidelberg 2010

"A term 'statistical dependence' rather than dependence should perhaps be used to empha-
size that we are dealing with new concepts specific for probability theory which may not
coincide with the daily meaning of this concept."

Various concepts and measures of statistical dependence have been introduced, including Pearson's correlation coefficient, Kendall's $\tau$ and Spearman's $\rho$ among others. These dependence measures are not in conformity with the above understanding
of dependence. Others, which can be broadly described as regression dependence
concepts, including Pearson's correlation ratio $\eta^2$, Goodman-Kruskal's $\tau$ and the
notion of positive and negative regression dependence (Lehmann, 1966), agree with
it. Literary use of correlation which covers any notion of dependence contributes to
creating frequent misconceptions about formalization of this concept.

A historical account of the dependence concepts mentioned above can be found
in Drouet Mari and Kotz (2001). Joe (1997) discussed using copulas in this context
and also other dependence concepts and measures as well as applications in reliability theory. See also Lai and Xie (2006), Szekli (1995) and the monograph edited by
Block, Sampson and Savits (1990). For other important contributions to the concept
of dependence see Tjøstheim (1996) and Skaug and Tjøstheim (1993).

The goal of the paper is to provide another look at regression approach to "statistical dependence" which is in line with our daily understanding of dependence.
In particular, we shall view statistical dependence as the quality or state of variables
being influenced by others, in accordance the interpretation given in the Meriam-
Webster dictionary. With this notion of dependence, we will introduce its measures
which quantify how the output variable varies with respect to changes in the input
variables. The novelty here is that the presence of all other variables influencing
output variable, summarily described as noise, is taken into account. In particular,
the proposed measures depend on strength of interaction between the input variables
and the noise in contrast e.g. to Pearson correlation ratio $\eta^2$. This approach has tight
connections with various areas including causal inference, nonlinear system theory,
prediction, experimental design, information theory and risk management. More
importantly, the proposed dependence measures provide a useful tool for statistical
analysis of complicated stochastic systems. They are a convenient vehicle for developing an asymptotic theory which is useful for the statistical inference, control,
prediction and other objectives.

The rest of the paper is structured as follows. In Section 2 we discuss in detail our
understanding of dependence and introduce various bivariate dependence measures,
as well as their connections with information theory. In Section 3 we shall mention
some connections with reliability theory regarding concepts of positive dependence.
Multivariate dependence is treated in Section 4. Section 5 presents moment inequalities and limit theorems for sums of random variables with complicated dependence
structures. We consider here a special type of causal processes being measurable
functions of Bernoulli shifts pertaining to independent and identically distributed
(iid) random variables. Some useful tools which are of independent interest are also
provided.

## 2 Bivariate dependence

For a pair of jointly distributed random variables $(X, Y)$, let $F_{XY}(x, y) = \mathbb{P}(X \leq x, Y \leq y)$, $x, y \in \mathbb{R}$, be the joint distribution function and $F_{Y|X}(y|x) = \mathbb{P}(Y \leq y|X = x)$ the conditional distribution function of $Y$ given $X = x$. For $u \in (0, 1)$, define the conditional quantile function

$$G(x, u) = \inf\{y \in \mathbb{R} : F_{Y|X}(y|x) \geq u\}. \tag{1}$$

By definition, for fixed $x$, $G(x, u)$ is a nondecreasing function in $u$. The conditional quantile function $G$ plays an important role in our theory of dependence. Let $U$ be a uniform$(0, 1)$ distributed random variable and assume that $U$ and $X$ are independent. Then the distributional equality

$$(X, Y) =_{\mathscr{D}} (X, G(X, U)) \tag{2}$$

holds. Based on (2), we can view $Y$ as the outcome of the bivariate function $G(\cdot, \cdot)$:

$$Y = G(X, U) \tag{3}$$

In other words, $Y$ is viewed as the output from a random physical system with $X$ and $U$ being the input and the noise or error, respectively. If we call $X$ an independent variable and $Y$ a dependent variable, then (3) becomes a regression problem if the goal is to find or estimate $G$. Observe that the right hand side of (1) can be written as a transformation of two *independent* uniform$(0, 1)$ distributed random variables $U$ and $V$, namely $(F_X^{-1}(V), G(F_X^{-1}(V), U))$. This is different from copula construction when $(X, Y)$ is transformed to a vector $(V_1, U_1) = (F_X(X), F_Y(Y))$ and dependence of $X$ and $Y$ is reflected by dependence of $U_1$ and $V_1$. These two frameworks have different ranges of applicability. Ours can lead to dependence measures that are useful for developing moment inequalities and limit theorems for stochastic processes (cf. Section 5).

*Remark 0.1.* If $F_{Y|X}(y|x)$ is strictly increasing and continuous in $y$, then it is obvious that $U := F_{Y|X}(Y|X)$ is uniform$(0, 1)$ distributed and it is independent of $X$. In this case we have a stronger version of (2): the identity $(X, Y) = (X, G(X, U))$ holds almost surely, while (2) holds only in the distributional sense. For a general construction yielding almost sure equality see Rüschendorf (1981).                    $\diamond$

With (3), we generically interpret dependence as how the output $Y$ depends on the input $X$ in *the presence of the noise* $U$. This interpretation seems to provide some clarifications of the troublesome confusion on the understanding of dependence mentioned in Mosteller and Tukey (1977, p. 262):

> We must be clearer about the abused word "dependence" and its relatives. When we say "y depends on x", sometimes we intend *exclusive dependence*, meaning that, if x is given, then the value of y follows, usually because the existence of a law is implied. In mathematics if y is the area of a circle and r the radius, then $y = \pi r^2$ illustrates this regular type of *exclusive dependence*.

At other times "y depends on x" means *failure of independence*, usually in the sense of "other things being equal" as in "the temperature of the hot water depends on how far the faucet is from the heater." Clearly, it may depend on other things, such as the setting of the heater and the building's temperature, to say nothing of whether the long pipe between the faucet and the heater is on a cold outside wall or a warm inside one. These are two quite different ideas of dependence, and the use of one word for both has often led to troublesome, if not dangerous, confusion.

Then there are the mathematical usages "dependent variable" and "independent variable". These have been extremely effective in producing confusion when dealing with data.

If $G(x,u)$ does not depend on $u$, then we can write $Y = g(X)$ and hence $Y$ functionally depends on $X$. Namely if $X = x$ is given, then the value of $Y$ is explicitly known. This is the exclusive dependence discussed above. On the other hand, if $G(x,u)$ does not depend on $x$, then $Y$ is functionally as well as statistically independent of $X$. For situations between this two extremes, we interpret the meaning of "$Y$ depends on $X$" as "the state of $Y$ being influenced by $X$", as indicated by the explanation of dependence in the Merriam-Webster dictionary. For example, in the second example, if we let $Y$ be the temperature of the hot water and $X$ the distance from the heater to the faucet. Then $Y$ is certainly influenced by other factors which are summarily described by $U$ in the definition above. With this understanding, one can have a unified look at the two quite different ideas of dependence and thus help clarify the "troublesome, if not dangerous, confusion".

Furthermore, our approach leads to dependence measures that are useful in developing an asymptotic theory for complicated stochastic systems. Certainly there are many ways to implement and quantify the degree of dependence of $Y$ on $X$. Here we consider two types: global dependence measures and local dependence measures, which are dealt with in Sections 2.1 and 2.2, respectively.

## 2.1 Global dependence measures

From now on assume that a copy of the bivariate random vector under consideration is given for which (3) holds. To define our global dependence measures, we need to apply the idea of coupling. This idea, which is frequently used in contemporary statistics, can be used in a natural way to measure the strength of dependence. Let $X'$ be an iid copy of $X$ and let $X$ and $X'$ be independent of $U$. Measures defined below are asymmetric and are meant to gauge how strongly $Y$ depends on $X$.

**Definition 0.1.** Let $Y = G(X,U) \in \mathcal{L}^p$, $p > 0$; let $Y' = G(X',U)$. Define $\delta_p(X,Y) = \|Y - Y'\|_p$, and, for $p \geq 1$, $\tau_p(X,Y) = \|Y - \mathbb{E}(Y|U)\|_p$.

In Definition 0.1, $\delta_p = \delta_p(X,Y)$ is defined under the principle of *ceteris paribus* (with other things being the same), a desirable property in the theory of design of experiments. Specifically, the random noises $U$ in $Y = G(X,U)$ and $Y' = G(X',U)$ are the same, and the only difference is $X$ and $X'$. If we interpret (3) as a physical system with the causality relationship of $X$ probabilistically causing $Y$ (Mellor, 1998), then we view $X$ and $Y$ as the input and the output, respectively, and thus $\delta_p$ measures the overall dependence of $Y$ on $X$. In this sense we call $\delta_p$ *physical* or *functional dependence measure*. In the early practice of statistics, the absence of causal relation has been linked with probability through the concept of independence and statistical inference has been developed as a method for inquiry into features of causes (see Glymour (2006) and Stigler (1986)). Here $\delta_p$ is meant to be a measure of causal relationship when a cause $(X)$ is understood as any quantity whose prior variation would result in the subsequent variation of another quantity $(Y)$.

In certain situations, especially in observational studies, it is not always possible to change $X$ and the causality relationship may not be observable. In such cases $\delta_p$ is only a mathematical object which may not be practically meaningful. Nonetheless, the measure $\delta_p$ is useful in developing an asymptotic theory. Different concept of causality has been introduced e.g. by Granger (1969) who defines that $X$ is causing $Y$ when we are able to predict $Y$ with a greater precision using the value of $X$ than using all other information our disposal other than $X$. Other approaches of describing complicated dependence patterns include graphical Markov models (see e.g. Cowell et al., 1999) and Rubin's framework (Rubin, 1990).

Observe that both introduced measures depend in a complicated way on $G$, the conditional quantile function of $Y$ given $X$. Under appropriate regularity conditions, based on a random sample $(X_i, Y_i)_{i=1}^{n}$, $G$ can be estimated non-parametrically and ensuing estimates $\widehat{\delta}_p(X,Y)$ and $\widehat{\tau}_p(X,Y)$ can then be computed or simulated. If $Y$ is known to depend on $X$ in a parametric way, then it suffices to estimate the unknown parameter and then $\widehat{\delta}_p(X,Y)$ and $\widehat{\tau}_p(X,Y)$ can be computed through simulation. In general $\delta_p(X,Y)$ and $\tau_p(X,Y)$ are not directly estimable. However, we will argue that they can be very useful in theoretical analysis of dependence.

The two dependence measures $\delta_p$ and $\tau_p$ are closely related; see Proposition 0.1 below. In the definition of $\tau_p$, the conditional expectation $\mathbb{E}(Y|U)$ smoothes out the impact of $X$ on $Y$. Thus $\tau_p$ is a distance between $Y$ and its analogue for which the impact of $X$ has been averaged. Both measures quantify the degree of global dependence of $Y = G(X,U)$, as a random function of $X$, on $X$. With this interpretation, our dependence measures are naturally asymmetric and the degree of dependence of $Y$ on $X$ and that of $X$ on $Y$ are different. There certainly exist other ways to measure the discrepancy between $Y$ and $Y'$ (or $\mathbb{E}(Y|U)$) by using e.g. different metrics.

**Proposition 0.1.** *We have for $p \geq 1$ that $\tau_p(X,Y) \leq \delta_p(X,Y) \leq 2\tau_p(X,Y)$.*

*Proof.* Write $\tau_p = \tau_p(X,Y)$ and $\delta_p = \delta_p(X,Y)$. Since $\mathbb{E}(Y|U) = \mathbb{E}(Y'|X,U)$, $Y - \mathbb{E}(Y|U) = \mathbb{E}(Y - Y'|X,U)$. By Jensen's inequality, $\tau_p \leq \delta_p$. Since $X, X', U$ are independent, $\mathbb{E}(G(X,U)|U) = \mathbb{E}(G(X',U)|U)$. Then

$$\delta_p \leq \|G(X,U) - \mathbb{E}(G(X,U)|U)\|_p + \|\mathbb{E}(G(X',U)|U) - G(X',U)\|_p = 2\tau_p.$$

$\diamondsuit$

**Proposition 0.2.** $\delta_p(X,Y)$ *and* $\tau_p(X,Y)$ *are strong measures of dependence in the sense that* $\delta_p = 0$ $(\tau_p = 0)$ *if and only if $X$ and $Y$ are independent.*

*Proof.* By Proposition 0.1, it is enough to check the property for $\tau_p$. Clearly, $\tau_p = 0$ if and only if $G(X,U) = \mathbb{E}(G(X,U)|U) \, P_X \times P_U$ -a.e. which implies that for any $0 < u < 1 \; F_{Y|X}^{-1}(u|x)$ does not depend on $x$. This is equivalent to independence of $X$ and $Y$.                                                                                            $\diamond$

Thus both measures satisfy one of Rényi postulates on measures of dependence. Observe, however, that they are both regression type measures and the symmetry postulate fails for them.

A particular important case is $p = 2$. In this case $\delta_2 = \sqrt{2}\tau_2$ as $G(X,U) - \mathbb{E}(G(X,U)|U)$ and $G(X',U) - \mathbb{E}(G(X',U)|U)$ are conditionally independent given $U$ with a conditional mean equal to 0.

Let us call $\mathbb{E}(Y|X)$ a main effect of $X$ on $Y$. We have the following decomposition due to orthogonality

$$\|Y - \mathbb{E}Y\|^2 = \|Y - \mathbb{E}(Y|X) - \mathbb{E}(Y|U) + \mathbb{E}Y\|^2$$
$$+ \|\mathbb{E}(Y|X) - \mathbb{E}Y\|^2 + \|\mathbb{E}(Y|U) - \mathbb{E}Y\|^2. \tag{4}$$

Thus the variability of $Y$ is decomposed into three terms: variability of two main effects of $U$ and $X$ on $Y$ and the variability of their interaction. Moreover, we have that the sum of two first terms in the above formula equals $\|Y - \mathbb{E}(Y|U)\|^2 = \tau_2^2$. Define Pearson correlation ratio (nonparametric R-squared) of arbitrary square integrable random variables $Z$ and $Y$ as $\eta_{Z,Y}^2 = \operatorname{var}(\mathbb{E}(Y|Z))/\operatorname{var}(Y)$ (cf e.g. Doksum and Samarov (1995)). Note that the numerator of $\eta_{Z,Y}^2$ equals $\operatorname{var}(Y) - \mathbb{E}\operatorname{var}(Y|Z)$ thus it measures relative decrease of variability of $Y$ given $Z$; other measures of dependence such as Goodman-Kruskal's $\tau$ for categorical data are constructed in the same vein. The following proposition reveals the relationship between $\tau_2$, $\operatorname{var}(Y)$ and $\eta_{U,Y}$.

**Proposition 0.3.** *We have*
   (i)    $\operatorname{var}(Y) = \tau_2^2 + \|\mathbb{E}(Y|U) - \mathbb{E}Y\|^2 = \tau_2^2 + \eta_{U,Y}^2 \times \operatorname{var}(Y);$
   (ii)    $\tau_2^2 \geq \eta_{X,Y}^2 \times \operatorname{var}(Y).$

Inequality (ii) follows by noting that $\mathbb{E}(Y - \mathbb{E}(Y|U))|X) = \mathbb{E}(Y|X) - \mathbb{E}Y$.

Observe that it follows from (4) that when we consider $\tau_2$ as the measure of dependence of $Y$ on $X$, we subtract from the overall variability of $Y$ only the variability of the main effect of $U$, leaving the interaction term. This corresponds to the idea of quantifying how the output $Y$ depends on the input $X$ *in the presence of the noise $U$.*

Let $\rho(X,Y) = \operatorname{cov}(X,Y)/\sqrt{\operatorname{var}(X)\operatorname{var}(Y)}$ be correlation coefficient of $X$ and $Y$ and observe that comparing distances of $Y - \mathbb{E}Y$ from its projections on the subspaces of measurable and linear functions of $U$ we have that $\rho^2(U,Y) \leq \eta_{U,Y}^2$ (Cramér, 1946). Moreover, using independence of $X$ and $U$ we have

$$\operatorname{cov}^2(X,Y) = \mathbb{E}^2(X - \mathbb{E}X)(Y - \mathbb{E}(Y|U)) \leq \operatorname{var}(X)\tau_2^2.$$

Thus we obtain

**Proposition 0.4.** *The following inequality holds:*

$$\rho^2(X,Y) \le \tau_2^2/\text{var}(Y) \le 1 - \rho^2(U,Y).$$

Let us compare now $\tau_2^2$ with $\eta_{X,Y}^2$.

*Example 2.1.* Consider the heteroscedastic regression model

$$Y = f(X) + \sigma(X)\varepsilon,$$

where $\varepsilon$ and $X$ are independent, $\mathbb{E}\varepsilon = 0$ and $\sigma(X) > 0$ $P_X$-a.e. Then we have

$$F_{Y|X}^{-1}(u|x) = f(X) + \sigma(X)F_\varepsilon^{-1}(u),$$

where $F_\varepsilon$ is the distribution function of $\varepsilon$ and $F_\varepsilon^{-1}$ is the quantile function. Thus

$$\tau_2^2 = \|Y - \mathbb{E}(Y|U)\|^2 = \|f(X) - \mathbb{E}f(X) + (\sigma(X) - \mathbb{E}\sigma(X))F_\varepsilon^{-1}(U)\|^2$$
$$= \text{var}(f(X)) + \text{var}(\sigma(X))\mathbb{E}\varepsilon^2.$$

Note that $\text{var}(Y) = \text{var}(f(X)) + \mathbb{E}(\sigma^2(X))\mathbb{E}\varepsilon^2$. Thus

$$\tau_2^2 = \text{var}(Y) - (\mathbb{E}\sigma(X))^2\mathbb{E}\varepsilon^2.$$

Consider the following special cases

(i) (additive model) Let $\sigma(X) \equiv 1$. Then $\tau_2 = \sigma_{f(X)}$ and $\tau_2^2 = \text{var}(Y) - \text{var}(\varepsilon)$. Note that this is also true if $\varepsilon \equiv 0$. In particular, for $(X,Y) \sim N(m_X, m_Y, \sigma_X, \sigma_Y, \rho)$ we have that $Y = m_X + \sigma_Y(\rho/\sigma_X(X - m_X) + (1 - \rho^2)^{1/2}\varepsilon)$ where $\varepsilon \sim N(0,1)$ is independent of $X$ and thus $\tau_2 = |\rho|\sigma_Y$.

(ii) (multiplicative model) Let $f(X) \equiv 0$. Then $\tau_2 = \sigma_{\sigma(X)}(\mathbb{E}\varepsilon^2)^{1/2}$.

Observe that it follows from the above examples that for the additive model $\eta_{X,Y}^2 = \tau_2^2/\text{var}(Y)$ thus $\eta_{X,Y}^2$ coincides with $\tau_2^2$ up to a positive constant whereas for the multiplicative model $Y = X\varepsilon$ with $\mathbb{E}\varepsilon = 0$ we have that $\eta_{X,Y}^2 = 0$ whereas $\tau_2 > 0$. Thus in the last case the interaction effect is present which is detected by $\tau_2$ but not by $\eta_{X,Y}^2$.

*Example 2.2.* (One way ANOVA). Assume that $Y = G(X,U)$ in (3) has the special form $Y = \mu_X + \varepsilon$, where $X$ is uniformly distributed over $\{1,2,\ldots,k\}$, and $\varepsilon = H(U)$ has finite second moment. Let $\bar{\mu} = \sum_{i=1}^n \mu_i/k$. Then $\mathbb{E}(Y|U) = \varepsilon + \bar{\mu}$, and $\tau_2^2 = k^{-1}\sum_{i=1}^n (\mu_i - \bar{\mu})^2$ corresponds to the treatment effect.

*Example 2.3.* Let $\widetilde{G}$ be a conditional quantile function corresponding to a distribution function (copula) of a transformed vector $(F_X(X), F_Y(Y))$. Then it is easy to check that $F_{F(Y)|F(X)}^{-1}(u|F(x)) = F_Y(F_{Y|X}^{-1}(u|x))$ and thus $\widetilde{G}(u,x) = F_Y(G(u, F_X^{-1}(x)))$. Moreover, the definition of $\delta_p$ for the pair $(F_X(X), F_Y(Y))$ is $\|\widetilde{G}(W,U) - \widetilde{G}(W',U)\|_p$, where $W, W', U$ are independent $U[0,1]$ random variables. Equivalently, $\delta_p = \|F_Y(Y) - F_Y(Y')\|_p$, where $Y = G(X,U)$, $Y' = G(X',U)$ and $X'$ is an independent copy of $X$.

*Example 2.4.* Let $X$ and $Y$ be $0/1$-valued and $\mathbb{P}(X = i, Y = j) = p_{ij}, i,j = 0,1$. Let $p_0. = p_{00} + p_{01}$ and $p_1. = p_{10} + p_{11}$. Then $p_0. + p_1. = 1$ and $\text{cov}(X,Y) = p_{11} - (p_{10} +$

$p_{11})(p_{01}+p_{11}) = p_{11}p_{00} - p_{01}p_{10}$. Let $G(X,U) = X\mathbf{1}_{U\leq p_{11}/p_{1.}} + (1-X)\mathbf{1}_{U\leq p_{01}/p_{0.}}$. Then (2) holds. Since $X$ and $U$ are independent and $\mathbb{E}X = p_{1.}$, we have

$$\tau_2^2(X,Y) = \|Y - \mathbb{E}(Y|U)\|_2^2 = \text{var}(X)\mathbb{E}|\mathbf{1}_{U\leq p_{11}/p_{1.}} - \mathbf{1}_{U\leq p_{01}/p_{0.}}|$$
$$= \text{var}(X)|p_{11}/p_{1.} - p_{01}/p_{0.}| = |p_{11}p_{00} - p_{01}p_{10}| = |\text{cov}(X,Y)|.$$

Observe that in this case we have symmetry: $\tau_2(X,Y) = \tau_2(Y,X)$.

## 2.2 Local dependence measures

In Definition 0.1 we introduced global dependence measures which quantify the difference between $Y = G(X,U)$ and its coupled version $Y' = G(X',U)$. In certain applications it is desirable to know the local dependence of $Y$ on $X$ varies with changing $x$. A natural way is to consider a properly normalized distance between $G(x,U)$ and $G(x+\delta,U)$, where $\delta \in \mathbb{R}$ is small.

**Definition 0.2.** Assume that the limit $G'(x,U) = \lim_{\delta\to 0} \delta^{-1}[G(x+\delta,U) - G(x,U)]$ exists almost surely. For $p > 0$ define $\delta_p(x) = \|G'(x,U)\|_p$ if the latter exists.

The local dependence measure $\delta_p(x)$ quantifies the dependence of $Y = G(X,U)$ on $X$ at $X = x$ in the presence of the noise $U$. If $G(x,u)$ does not depend on $u$ i.e. $Y$ is a function of $X$, we write $G(x,u) = G(x)$ and then $G'(x,U)$ is just the derivative $G'(x)$. The following proposition shows that the global dependence measure $\delta_p$ is bounded by a $\delta_p(x)$-weighted distance between $X$ and $X'$.

**Proposition 0.5.** Let $p \geq 1$. Then $\delta_p \leq \|\int_X^{X'} \delta_p(x)dx\|_p$.

*Proof.* The inequality trivially holds if $p = 1$. For $p > 1$ let $q = p/(p-1)$ and $\lambda(z) = \delta_p^{-1/q}(z)$. Then $q^{-1} + p^{-1} = 1$ and by Hölder's inequality,

$$|G(x,U) - G(x',U)| = \left|\int_x^{x'} G'(z,U)dz\right|$$
$$\leq \left|\int_x^{x'} |G'(z,U)|^p \lambda^p(z)dz\right|^{1/p} \times \left|\int_x^{x'} \lambda(z)^{-q}dz\right|^{1/q}.$$

Then $\mathbb{E}[|G(x,U) - G(x',U)|^p] \leq |\int_x^{x'} \delta_p(z)dz|^p$ and the proposition follows. ◇

In Definition 0.2, $\delta_p(x)$ involves the Euclidean distance between $G(x,U)$ and $G(x+\delta,U)$. An alternative way is to use the information divergence. Let $p(\cdot|x)$ be the density function of $G(x,U)$. There exist many ways to measure the distance between $p(\cdot|x)$ and $p(\cdot|x+\delta)$.

**Definition 0.3.** (i) $I_1$ local dependence measure. Assume that as $t \to 0$, one has the approximation

$$\int [p(y|x+t) - p(y|x)] \log \frac{p(y|x+t)}{p(y|x)} dy = t^2 I_1(x) + o(t^2), \tag{5}$$

where the left hand side of (5) is symmetrized Kullback-Leibler divergence between $p(\cdot|x)$ and $p(\cdot|x+t)$ and

$$I_1(x) = \int \frac{[\partial p(y|x)/\partial x]^2}{p(y|x)} dy \tag{6}$$

Note that $I_1(x)$ may be viewed as Fisher's information of $p(\cdot|x)$ when $x$ is treated as its parameter.

(ii) $I_2$ local dependence measure. Assume that as $t \to 0$, one has the approximation

$$\int [p(y|x+t) - p(y|x)]^2 dy = t^2 I_2(x) + o(t^2), \tag{7}$$

where

$$I_2(x) = \int [\partial p(y|x)/\partial x]^2 dy. \tag{8}$$

Observe that equation (5) is satisfied under regularity conditions in view of equality $\log(1+\Delta) = \Delta + \mathcal{O}(\Delta^2)$ with $\Delta = (p(y|x+t) - p(y|x))/p(y|x)$.

In the context of nonlinear prediction theory, Fan and Yao (2003) introduced local measures $I_1$ and $I_2$ to quantify sensitivity to initial values. Wu (2008) applied $I_2(x)$ to study weak convergence of empirical processes of dependent random variables. Blyth (1994) noticed that $I_1(x)$ has a nice property: $I_1(x) \geq [g'(x)]^2/\text{var}(Y|X = x)$, where $g(x) = \mathbb{E}(Y|X = x) = \mathbb{E}G(x, U)$ and $\text{var}(Y|X = x) = \mathbb{E}(Y^2|X = x) - g^2(x)$. The last inequality is just the conditional version of the Cramér-Rao inequality.

Hellinger distance is another widely used metric to measure distances between distributions. Proposition 0.6 shows that the local Hellinger dependence measure is closely related to $I_1(x)$ and they are equivalent in the local sense. It easily follows from the Lebesgue dominated convergence theorem.

**Proposition 0.6.** *Assume that $\frac{\partial}{\partial x}\sqrt{p(y|x)}$ exists for $y \in \mathbb{R}$ and for some $g_x(\cdot) \in \mathcal{L}^2(\mathbb{R})$ it holds that $|\sqrt{p(y|x+t)} - \sqrt{p(y|x)}| \leq g_x(y)|t|$ for all $y \in \mathbb{R}$ and sufficiently small $t$. Then we have*

$$\lim_{t \to 0} \frac{1}{t^2} \int [\sqrt{p(y|x+t)} - \sqrt{p(y|x)}]^2 dy = \frac{I_1^2(x)}{4}. \tag{9}$$

A special class of local divergencies $\mathscr{D}$ such that $\mathscr{D}(p(y|x+t), p(y|x)) \sim t^2 I_1(x)$ when $t \to 0$ is considered in Blyth (1994). It is proved that for such measures, called local Rao divergencies, we have

$$\lim_{t \to 0} \frac{\mathscr{D}(p(y|x+t), p(y|x))}{t^2} \geq \frac{g'^2(x)}{\text{var}(Y|X = x)}.$$

*Example 2.5.* Consider the special case $G(x,u) = g(x) + H(u)$, where $H(U)$ has density function $h$ and $g$ is differentiable. Then $p(y|x) = h(y - g(x))$ and $\partial p(y|x)/\partial x = h'(y - g(x))g'(x)$. Let $I_h = \int [h'(y)]^2/h(y)dy$ be the Fisher information pertaining to $h$. Then by (6), the Kullback-Leibler local dependence measure $I_1(x) = [g'(x)]^2 I_h$.

*Example 2.6.* Consider again the heteroscedastic regression model. As $G(x,u) = f(x) + \sigma(x)F_\varepsilon^{-1}(u)$ we have

$$\delta_2^2(x) = \int_0^1 (f'(x) + \sigma'(x)F_\varepsilon^{-1}(u))^2 \, du$$
$$= f'^2(x) + \sigma'^2(x)\int (F_\varepsilon^{-1}(u))^2 \, du + 2f'(x)\sigma'(x)\int F_\varepsilon^{-1}(u) \, du$$
$$= f'^2(x) + \sigma'^2(x)\text{var}(\varepsilon).$$

Let $g(x) = \mathbb{E}(Y|X = x)$ and observe that the derivative $g'(x)$ indicates how the conditional expected value of $Y$ given $X = x$ changes with changing $x$. Bjerve and Doksum (1993) introduced correlation curve $\rho(x)$, still another measure of local dependence, which is based on $g'(x)$ and takes the form

$$\rho(x) = \frac{\sigma_1 g'(x)}{(g'(x)^2\sigma_1^2 + \sigma^2(x))^{1/2}},$$

where $\sigma^2(x) = \text{var}(Y|X = x)$ and $\sigma_1^2 = \text{var}(X)$. This is a local analogue of equality holding for a correlation coefficient of bivariate normal vector $\rho = \beta\sigma_1/(\beta^2\sigma_1^2 + \sigma^2)^{1/2}$ with $\beta$ denoting slope of $Y$ regressed on $X$ and $\sigma^2 = \text{var}(Y|X)$. Correlation curve $\rho(x)$ is easily estimated by means of local linear smoothers which yield natural estimator of $g'(x)$. Observe, however, that for multiplicative model $\rho(x) \equiv 0$. This is due to the fact that correlation curve measures strength of locally *linear* relationship. In order to remedy this drawback Doksum and Froda (2000) consider correlation ratio given $X$ belongs to a small neighborhood of $x$ calibrated to coincide with $\rho^2$ in linear models. The following proposition relates $\delta_p(x)$ with numerator of the ratio defining correlation curve.

**Proposition 0.7.** *Let* $G_\eta(x,u) = \sup_{|x-y|\le\eta} |\partial G(y,u)/\partial y|$. *Assume that for some* $\eta > 0$, *we have* $G_\eta(x,\cdot) \in \mathscr{L}^p$ *for some* $p \ge 1$. *Then* $|g'(x)| \le \delta_p(x)$.

*Proof.* Define $D_\delta(u) = [G(x+\delta,u) - G(x,u)]/\delta$. Then

$$\left|\lim_{\delta\to 0} \frac{g(x+\delta) - g(x)}{\delta}\right| = \left|\lim_{\delta\to 0}\int_0^1 D_\delta(u)\,du\right| \le$$
$$\lim_{\delta\to 0}\left(\int_0^1 |D_\delta(u)|^p\,du\right)^{1/p} = \left(\int_0^1 \lim_{\delta\to 0}|D_\delta(u)|^p\,du\right)^{1/p},$$

where the last equality follows from the Lebesgue dominated convergence theorem. Thus the proposition follows.                                                                                                     ◇

Observe that for Example 2.6 the proposition is trivially satisfied as $g(x) = f(x)$.

## 3 Connections with reliability theory

In reliability theory the notion of dependence has been widely studied and various concepts of positive dependence have been proposed; see Lehmann (1966) and the recent exposition by Lai and Xie (2006). Here we shall discuss how our concept of dependence is related to those in reliability theory.

Note that from the definition of $G$ in (1), $G(x, \cdot)$ is nondecreasing for any fixed $x$. Imposing certain conditions on $G(\cdot, u)$ amounts to assumptions on the character of dependence. In particular, it follows immediately from the definition of $G$ that a property that $G(\cdot, u)$ is nondecreasing is equivalent to positive regression dependence (PRD) introduced in Lehmann (1966), for which family of conditional distributions $F_{Y|X}(\cdot|x)$ are ordered w.r.t. $x$. This property is also called $Y$ stochastically increasing on $X$ (SI) in Barlow and Proschan (1975).

**Proposition 0.8.** $G(\cdot, u)$ is nondecreasing for any $u$ if and only if $\mathbb{P}(Y > y | X = \cdot)$ is nondecreasing for any $y$.

Define $G^{-1}(y|x) = \inf\{u : G(x,u) \geq y\}$. Obviously, in the case when $F_{Y|X}(\cdot|x)$ is continuous and strictly increasing for any $x$ then $G^{-1}(y|x) = F_{Y|X}(y|x)$. Define $G_{x'x}(u) = G^{-1}(G(x,u)|x')$. Then positive regression dependence is equivalent to $G_{x'x}(u) \leq u$ for any $x' > x$ (Capéraà and Genest, 1990).

Note that the above obvious equivalence of Proposition 0.8 yields a straightforward proof that positive regression dependence implies association i.e. that

$$\mathrm{Cov}(f_1(X,Y), f_2(X,Y)) \geq 0 \tag{10}$$

holds for any $f_1, f_2$ which are coordinate-wise nondecreasing and such that the covariance exists. This has been proved first by Esary et al. (1967). Namely, observe that the functions $h_i(x,u) = f_i(x, G(x,u))$, $i = 1, 2$, are nondecreasing in both arguments. Thus

$$\mathbb{E}(h_1(X,U') - h_1(X',U'))(h_2(X,U) - h_2(X',U)) \geq 0$$

and

$$\mathbb{E}(h_1(X,U) - h_1(X,U'))(h_2(X,U) - h_2(X,U')) \geq 0,$$

where $(X', U')$ is an independent copy of $(X, U)$. Summing both inequalities we obtain $2\mathrm{cov}(h_1(X,U), h_2(X,U)) \geq 0$. The property has natural generalization to multivariate case which we will discuss later. Lehmann (1966) has shown that PRD implies that $\mathbb{P}(Y > y | X \leq \cdot)$ is nondecreasing, whereas the fact that $\mathbb{P}(Y > y | X > \cdot)$ is nondecreasing implies association of $X$ and $Y$ (Esary and Proschan, 1972).

In particular, association implies $\mathrm{cov}(f(X), g(Y)) \geq 0$ for any nondecreasing $f$ and $g$. This is easily seen to imply positive quadrant dependence (PQD) condition $\mathbb{P}(X > x, Y > y) \geq \mathbb{P}(X > x)\mathbb{P}(Y > y)$ for $x, y \in \mathbb{R}$. The following generalization is possible. If $(X_i, Y_i), i = 1, 2, \ldots, n$ are independent copies of $(X, G(X, U))$ with

$G(\cdot, u)$ nondecreasing then it follows from Lehmann (1966) that if $r, s : \mathbb{R}^n \to \mathbb{R}$ are concordant functions such that direction of monotonicity for each univariate argument is the same then $\text{cov}(r(\mathbf{X}), s(\mathbf{Y})) \geq 0$, where $\mathbf{X} = (X_1, X_2, \ldots, X_n)$ and $\mathbf{Y} = (Y_1, Y_2, \ldots, Y_n)$.

Observe also that for nondecreasing $G(\cdot, u)$, $Y = G(X, U)$ is a monotone mixture with $X$ (c.f. Jogdeo (1978)) meaning that the function $h(x) := \mathbb{E}(f(G(X, U))|X = x)$ is nondecreasing for any nondecreasing function $f$.

A pair $(X, Y)$ is likelihood ratio dependent (LRD) if its bivariate density $f(x, y)$ is totally positive of order 2 i.e. for any $x < x'$ and $y < y'$, $f(x, y)f(x', y') \geq f(x, y')f(x', y)$. Then provided that $F_{Y|X}(\cdot|x)$ is continuous and strictly increasing, LRD is equivalent to convexity of $G_{x'x}$ for any $x < x'$, where $G_{x'x}$ is the function defined above in terms of the function $G$.

We now discuss how some multivariate extensions of these concepts of positive dependence can be interpreted in the presented framework. Note that in the construction of function $G$ one can assume that $\mathbf{X} = (X_1, X_2, \ldots, X_m)$ is a multivariate vector. Thus we have the representation $(\mathbf{X}, Y) =_{\mathscr{D}} (\mathbf{X}, G(\mathbf{X}, U))$. Another sequential construction is discussed in the next section. We now prove the extension of (10) to the case when $G(\mathbf{x}, u)$ is nondecreasing in any coordinate of $\mathbf{x}$.

**Proposition 0.9.** *Assume that* $\mathbf{X}$ *is associated and* $G(\mathbf{x}, u)$ *is nondecreasing in any coordinate of* $\mathbf{x}$. *Then* $(\mathbf{X}, G(\mathbf{X}, U))$ *is associated.*

*Proof.* Define $h_i(\mathbf{x}, u) = f_i(\mathbf{x}, G(\mathbf{x}, u))$ for $i = 1, 2$. Then

$$\mathbb{E}\{(h_1(\mathbf{X}, U) - h_1(\mathbf{X}, U'))(h_2(\mathbf{X}', U) - h_2(\mathbf{X}', U'))\} \geq 0.$$

Moreover, for any fixed $u$ as $h_i(\mathbf{x}, u)$ are nondecreasing in each coordinate of $\mathbf{x}$ we have

$$0 \leq 2\text{cov}(h_1(\mathbf{X}, u), h_2(\mathbf{X}, u)) = \mathbb{E}\{(h_1(\mathbf{X}, u) - h_1(\mathbf{X}', u))(h_2(\mathbf{X}, u)) - h_2(\mathbf{X}', u))\}$$

since $\mathbf{X}$ is associated. Integrating the second inequality w.r.t. distribution of $U$ and adding the result to the first inequality we get $2\text{cov}(h_1(\mathbf{X}, U), h_2(\mathbf{X}, U)) \geq 0$ which proves the assertion.                                                                                      $\diamondsuit$

Proposition 0.9 is a generalization of (10) by noting that a univariate variable $X$ is always associated and thus $(X, G(X, U))$ is associated for nondecreasing $G(\cdot, u)$.

Consider the following generalization of positive regression dependence. The random vector $\mathbf{X}$ is conditionally increasing in sequence (CIS) if $\mathbb{P}(X_i > x|\mathbf{X}_{i-1} = \mathbf{x}_{i-1})$ is increasing in $x_1, \ldots, x_{i-1}$ for any $i$, where $\mathbf{X}_{i-1} = (X_1, \ldots, X_{i-1})$ and $\mathbf{x}_{i-1} = (x_1, \ldots, x_{i-1})$. Define $G_i$ as the function $G$ for $\mathbf{X} = \mathbf{X}_{i-1}$ and $Y = X_i$ for $i = 2, \ldots, n$. Then CIS is equivalent to $G_i(\mathbf{x}, u)$ being nondecreasing in each coordinate of $\mathbf{x}$, $i = 2, \ldots, n$. Then it follows easily by induction from Proposition 0.9 that CIS implies positive association (for a different proof see Barlow and Proschan (1975)). Also the stronger property of $\mathbf{X}$ being conditionally increasing (CI) (see e.g. Drouet Mari and Kotz (2005), p. 39) can be expressed by monotonicity of appropriately defined functions $G$.

## 4 Multivariate dependence

When analyzing complicated stochastic systems, one has to deal with multivariate dependence. To this end, we can introduce a multivariate version of (3). Earlier related discussions are given in Rosenblatt (1952). Consider the random vector $(X_1, \ldots, X_n)$. Let $\mathbf{X}_m = (X_1, \ldots, X_m)$. As in (2), there exists a measurable function $G_n$ and a standard uniform random variable $U_n$, independent of $\mathbf{X}_{n-1}$, such that

$$\mathbf{X}_n =_{\mathscr{D}} (\mathbf{X}_{n-1}, G_n(\mathbf{X}_{n-1}, U_n)). \tag{11}$$

Here $G_n$ is the conditional quantile of $X_n$ given $\mathbf{X}_{n-1} = (X_1, \ldots, X_{n-1})$. In the theory of risk management, $G_n(\mathbf{X}_{n-1}, u)$ is the value-at-risk (VaR) at level $u$ [cf. J. P. Morgan (1996)]. Namely, given $\mathbf{X}_{n-1}$ which represents the information available up to time $n-1$, $G_n(\mathbf{X}_{n-1}, u)$ is the threshold or risk level at which the probability of $X_n$ exceeds is $u$. Let $\mathbf{U}_m = (U_1, \ldots, U_m)$. Dependence structure of $(X_1, X_2, \ldots, X_n)$ is essential in determining VaRs. For a readable introduction see Embrechts et al. (2002).

Iterating (11), we have

$$\begin{pmatrix} X_1 \\ X_2 \\ \ldots \\ X_n \end{pmatrix} =_{\mathscr{D}} \begin{pmatrix} X_1 \\ G_2(\mathbf{X}_1, U_2) \\ \ldots \\ G_n(\mathbf{X}_{n-1}, U_n) \end{pmatrix} =_{\mathscr{D}} \begin{pmatrix} H_1(\mathbf{U}_1) \\ H_2(\mathbf{U}_2) \\ \ldots \\ H_n(\mathbf{U}_n) \end{pmatrix}, \tag{12}$$

where $H_1, \ldots, H_n$ are measurable functions. The above distributional equality serves as a standard method of simulating multivariate distributions (see e.g. Deák (1990), chapter 5) and the pertaining method is called the standard construction. The above representation is used in Rüschendorf and de Valk (1993) to characterize some subclasses of Markov sequences. Note that for (12) no regularity of distribution of $\mathbf{X}$ is needed in contrast to the corresponding construction yielding $F(\mathbf{X}) = \mathbf{U}_n$ almost surely for which absolute continuity of distribution of $\mathbf{X}$ is usually assumed. For other uses of the standard construction see e.g. Arjas and Lehtonen (1978) and Li et al. (1996). If $(X_i)$ is a Markov chain, then in view of definition of $G_i$ as a conditional quantile $G_i(\mathbf{X}_{i-1}, U_i)$ in (12) can be written as a function of $X_{i-1}$. Denote by $H_j(\mathbf{Y}_{j-1}, u)$, $j = 2, \ldots, n$ the analogue of the function $G_j$ defined in (12) for the vector $\mathbf{Y}_n = (Y_1, Y_2, \ldots, Y_n)$. It is proved in Arjas and Lahtonen (1978) that if $X_1$ is stochastically smaller than $Y_1$ and $G_i(\mathbf{x}_{i-1}, u) \leq H_i(\mathbf{y}_{i-1}, u)$ for any $\mathbf{x}_{i-1} \leq \mathbf{y}_{i-1}$ then there exist copies of $\widetilde{\mathbf{X}}_n$ and $\widetilde{\mathbf{Y}}_n$ of $\mathbf{X}_n$ and $\mathbf{Y}_n$ respectively, such that $\widetilde{\mathbf{X}}_n \leq \widetilde{\mathbf{Y}}_n$ almost surely. This is equivalent to $\mathbb{E}f(\mathbf{X}) \leq \mathbb{E}f(\mathbf{Y})$ for any nondecreasing $f$ (cf e.g. Liggett (1985)).

In the sequel we shall discuss the problem whether elements of an infinite sequence $(X_i)_{i \in \mathbb{Z}}$ can be expressed as functions of iid random variables, namely whether an analogue of (12) with infinitely many rows holds. With such representations, one can establish limit theorems and moment inequalities under simple and easily workable conditions; see Section 5. The latter problem belongs to the stochastic realization theory (Borkar, 1993). Wiener (1958) first considered this representation problem for stationary and ergodic processes. As a special case, if $(X_i)$ is a Markov chain with the form $X_i = G_i(X_{i-1}, U_i)$, Theorem 4.1 asserts that, under some regularity conditions there exists a copy $\widetilde{X}_i$ of $X_i$ such that $(\widetilde{X}_i)_{i \in \mathbb{Z}} =_{\mathscr{D}} (X_i)_{i \in \mathbb{Z}}$ and $\widetilde{X}_i$ is expressed as $H_i(\ldots, U_{i-1}, U_i)$, a function of iid random variables.

**Theorem 4.1.** *Assume that $(X_i)$ satisfies the recursion*

$$X_i = G_i(X_{i-1}, U_i) =: F_i(X_{i-1}), \ i \in \mathbb{Z}, \tag{13}$$

*where $U_i$ are iid standard uniform random variables. Here $F_i$ are independent random maps $F_i(x) = G_i(x, U_i)$. Assume that for some $\alpha > 0$ we have*

$$\sup_{i \in \mathbb{Z}} L_i < 1, \ where \ L_i = \sup_{x \neq y} \frac{\|G_i(x, U) - G_i(y, U)\|_\alpha}{|x - y|} \tag{14}$$

*and for some $x_0$,*

$$\sup_{i \in \mathbb{Z}} \|G_i(x_0, U)\|_\alpha < \infty. \tag{15}$$

*Then the backward iteration $F_i \circ F_{i-1} \circ F_{i-2} \ldots$ converges almost surely and the limit forms a non-stationary Markov chain which is a solution to (13).*

*Proof.* The theorem can be proved along the same line as Theorem 2 in Wu and Shao (2004), where the stationary case with $G_i \equiv G$ is dealt with. Their argument improves the classical result by Diaconis and Freedman (1999) and it can be easily generalized to the current non-stationary setting. Here we provide an outline of the argument and omit the details. Let $\mathscr{X}$ be the state space of $X_i$. For $i \geq j$ define the random map $F_{i,j}$ by

$$F_{i,j}(x) = F_i \circ F_{i-1} \circ \ldots \circ F_j(x), \quad x \in \mathscr{X},$$

where $\circ$ denotes function composition. Let $q = \min(1, \alpha)$. By (14) and (15), there exists a $\rho \in (0, 1)$ such that $\mathbb{E}\{|F_{i,j}(x_0) - F_{i,j+1}(x_0)|^q\} = O(\rho^{i-j})$. Hence $\sum_{j=-\infty}^{i} |F_{i,j}(x_0) - F_{i,j+1}(x_0)|^q < \infty$ almost surely and the sequence $F_{i,j}(x_0)$, $j = i, i-1, i-2, \ldots$, has an almost sure limit $Z_i$ (say). One can show that $Z_i \in \mathscr{L}^\alpha$. Also, following the argument of Theorem 2 in Wu and Shao (2004), we have for any $x$, the sequence $F_{i,j}(x)$, $j = i, i-1, i-2, \ldots$, has the same limit $Z_i$ which does not depend on $x$. Let $X_i = Z_i$, then (13) holds.                                      $\diamond$

Next we consider the problem of whether stationary and ergodic processes can be expressed as functionals of iid random variables. Let $(X_i)_{i \in \mathbb{Z}}$ be a stationary and ergodic process. Following (12), one may think whether it can be generalized to the case of infinite sequences, namely whether one can find iid standard uniform random variables $U_j$ and a measurable function $H$ (recalling stationarity) such that

$$(X_i)_{i \in \mathbb{Z}} =_{\mathscr{D}} (H(\ldots, U_{i-1}, U_i))_{i \in \mathbb{Z}}. \tag{16}$$

Intuitively, using $(X_i)_{i \in \mathbb{Z}} = (G(\mathbf{X}_{i-1}, U_i))_{i \in \mathbb{Z}}$, one can recursively write

$$X_i =_{\mathscr{D}} G(\mathbf{X}_{i-1}, U_i) =_{\mathscr{D}} G_2(\mathbf{X}_{i-2}; U_{i-1}, U_i) =_{\mathscr{D}} G_3(\mathbf{X}_{i-3}; U_{i-2}, U_{i-1}, U_i) = \ldots (17)$$

A major difficulty here is that the sequence $G_j(\cdot; U_{i-j+1}, \ldots, U_{i-1}, U_i)$ may not have a limit. In Theorem 4.1, due to the special Markovian structure and conditions (14) and (15), it is proved that the latter sequence does converge. In general the convergence is not guaranteed. Wiener (1958) argued that (16) should hold in a stronger sense that the distributional equality is an equality. Rosenblatt (1959) showed that Wiener's result is not true. See also Rosenblatt (1971), Kallianpur (1981) and Tong (1990) for more details. D.Volny (personal communications) pointed out that (16) is generally not valid. Processes of the form $(H(\ldots, U_{i-1}, U_i))_{i \in \mathbb{Z}}$ are necessarily Bernoulli (Ornstein, 1973). It is known that some stationary and ergodic processes are not Bernoulli (see Ornstein (1973) and Kalikow (1982)). One such example is given in Rosenblatt (2009).

## 5 Moment inequalities and limit theorems

Let $(X_i)$ be a stochastic process. Knowledge of $S_n = X_1 + \ldots + X_n$ is of critical importance in the study of random processes and various dependence measures for stochastic processes are relevant tools for this purpose. We mention construction of various mixing coefficients (see e.g. Bradley (2005)), weak dependence coefficients discussed in Dedecker et al. (2007), projective measures of dependence (Gordin (1969)) and measures based on mutual information of adjacent blocks of observations such as an excess entropy (Crutchfeld and Feldman (2003)). Moreover, special types of dependence, such as association, are useful for studying asymptotic properties of the partial sums (see e.g. Newman (1984)). Doukhan and Louhichi (1999) proposed an important dependence measure which concerns the decay rates of covariances between functionals of the past and the future of the underlying processes.

Here we shall study $S_n$ for causal processes by generalizing dependence measures given in Definition 0.1 to stochastic processes. Let $\varepsilon_i, i \in \mathbb{Z}$, be iid random variables and let $\xi_i = (\ldots, \varepsilon_{i-1}, \varepsilon_i)$. Consider the infinite stochastic system

$$X_i = H_i(\xi_i), \quad i \in \mathbb{Z}, \tag{18}$$

where $H_i$ are measurable functions. We view $\xi_i$ as the input and $X_i$ as the output of the system. If $H_i$ does not depend on $i$, then the process $(X_i)$ generated by (18) is stationary. Here we are interested in the behavior of $S_n = X_1 + \ldots + X_n$, which is helpful in understanding properties of the underlying stochastic system.

It is usually difficult to know or derive the close form of the distribution of $S_n$. For large values of $n$ asymptotic theory provides a useful approximation. To this end we shall introduce functional and predictive dependence coefficients. They are defined for a certain class of causal processes defined in (18), where $(\varepsilon_i)$ is an iid sequence. Let $(\varepsilon_i')_{i \in \mathbb{Z}}$ be an iid copy of $(\varepsilon_i)_{i \in \mathbb{Z}}$ and $\xi_{i,j} = (\ldots, \varepsilon_{i-1}^*, \varepsilon_i^*)$, where $\varepsilon_l^* = \varepsilon_l$ if $l \neq j$

and $\varepsilon_j^* = \varepsilon_j'$. Namely $\xi_{i,j}$ is a coupled version of $\xi_i$ with $\varepsilon_j$ in the latter replaced by $\varepsilon_j'$. For $k \geq 0$ and $p \geq 1$ define

$$\delta_{k,p} = \sup_{i \in \mathbb{Z}} \|H_i(\xi_i) - H_i(\xi_{i,i-k})\|_p \tag{19}$$

and, let the projection operator $\mathscr{P}_j(\cdot) = \mathbb{E}(\cdot|\xi_j) - \mathbb{E}(\cdot|\xi_{j-1})$,

$$\theta_{k,p} = \sup_{i \in \mathbb{Z}} \|\mathscr{P}_{i-k}H_i(\xi_i)\|_p = \sup_{i \in \mathbb{Z}} \|\mathbb{E}[H_i(\xi_i)|\xi_{i-k}] - \mathbb{E}[H_i(\xi_{i,i-k})|\xi_{i-k}]\|_p. \tag{20}$$

The second equality in (20) is due to the fact that:
$\mathbb{E}[H_i(\xi_i)|\xi_{i-k-1}] = \mathbb{E}[H_i(\xi_{i,i-k})|\xi_{i-k-1}] = \mathbb{E}[H_i(\xi_{i,i-k})|\xi_{i-k}]$. We observe that $\delta_{k,p}$ equals $\sup_{i \in \mathbb{Z}} \delta_p(\varepsilon_{i-k}, H_i(\xi_i))$ defined in Definition 0.1 i.e. $H_i(\xi_i)$ plays a role of $Y$ and $\varepsilon_{i-k}$ a role of $X$. Thus the strength of dependence of the random variable $X_i = H_i(\xi_i)$ observed at moment $i$ on the innovation $\varepsilon_{i-k}$ is quantified. Note that the counterpart of $\tau_p$ would be $\tau_{k,p} = \sup_{i \in \mathbb{Z}} \|H_i(\xi_i) - \mathbb{E}(H_i(\xi_i)|\xi_{i-k})\|_p$. Wu (2005) introduced functional and predictive dependence coefficients for stationary processes. Note that in (19) we assume that $\varepsilon_i, \varepsilon_j', i, j \in \mathbb{Z}$, are iid. It is meaningless if the independence assumption is violated. Under the construction (12), they are automatically iid uniform(0,1) random variables.

The following theorem concerns moments of $S_n$ and can be proved along the lines of Wu (2007). We refer to Gordin and Lifsic (1978) and Hannan (1979) who used a representation of a random variable as a sum of such projections to derive asymptotic results. A similar approach was later used by several authors, see Woodroofe (1992), Volny (1993), Dedecker and Merlevède (2002), Wu and Woodroofe (2004), Bryk and Mielniczuk (2005) and Merlevède and Peligrad (2006) among others.

**Theorem 5.1.** *Let* $p > 1$, $p' = \min(2, p)$ *and* $B_p = 18p^{3/2}/\sqrt{p-1}$. *Assume that*

$$\Theta_{0,p} := \sum_{i=0}^{\infty} \theta_{i,p} < \infty. \tag{21}$$

*For* $k \geq 0$ *let* $\Theta_{k,p} = \sum_{i=k}^{\infty} \theta_{i,p}$. *Then we have (i)* $D_j = \sum_{i=j}^{\infty} \mathscr{P}_j X_i$, $j \in \mathbb{Z}$, *form martingale differences with respect to* $\xi_j$ *and* $M_n = \sum_{i=1}^{n} D_i$ *satisfies*

$$\|S_n - M_n\|_p^{p'} \leq 3B_p^{p'} \sum_{j=1}^{n} \Theta_{k,p}^{p'}, \tag{22}$$

*and (ii)* $S_n^* = \max_{j \leq n} |S_j|$ *satisfies* $\|S_n^*\|_p \leq C_p n^{1/p'} \Theta_{0,p}$, *where* $C_p = pB_p/(p-1)$.

Theorem 5.1(ii) is closely related to the Efron-Stein inequality. Consider the special case $p = 2$ and let $X = G(\varepsilon_1, \varepsilon_2, \ldots, \varepsilon_n)$ be a function of independent $\varepsilon_1, \varepsilon_2, \ldots, \varepsilon_n$ and $X_i = G(\varepsilon_1, \varepsilon_2, \ldots, \varepsilon_i', \ldots, \varepsilon_n)$,. Then Efron-Stein inequality states that $\text{var}(X) \leq 2^{-1} \sum_{i=1}^{n} \mathbb{E}(X - X_i)^2$ (Efron and Stein, 1981).

**Theorem 5.2.** *Assume that* $X_i \in \mathscr{L}^p$, $2 < p \leq 4$, *and* $\mathbb{E}(X_i) = 0$. *Further assume that*

$$\sum_{k=1}^{\infty} (\delta_{k,p} + k\theta_{k,p}) < \infty. \tag{23}$$

Let $\sigma_i = \|D_i\|$, where $D_j = \sum_{i=j}^{\infty} \mathscr{P}_j X_i$. Then on a richer probability space, there exists a standard Brownian motion $\mathscr{B}$ such that

$$\max_{i \leq n} |S_i - \mathscr{B}(\sigma_1^2 + \ldots + \sigma_i^2)| = o_{\text{a.s.}} [n^{1/p} (\log n)^{1/2+1/p} (\log \log n)^{2/p}]. \tag{24}$$

The approximation (24) suggests that $S_i$ can be "regularized" by a Gaussian process $Z_i = \mathscr{B}(\sigma_1^2 + \ldots + \sigma_i^2)$ which has independent but non-stationary increments $Z_i - Z_{i-1}$. If $X_i$ are iid with $X_i \in \mathscr{L}^p$, then the classical Hungarian strong embedding (strong invariance principle) asserts that $\max_{i \leq n} |S_i - \mathscr{B}(i\sigma^2)| = o_{\text{a.s.}}(n^{1/p})$, where $\sigma = \|X_i\|$. For stationary processes Wu (2007) obtained (24). A careful check of the proof in the last paper shows that (24) also hold under condition (23). The details are omitted.

**Acknowledgements**  We would like to thank a referee for her/his many useful comments.

# References

[1] Arjas, E. and Lehtonen, T. (1978) Approximating many server queues by means of single server queues. *Math. Operation Research* **3**, 205-223.

[2] Barlow, R. and Proschan, F. (1975) *Statistical theory of reliability and life testing*, Rinehart and Winston, New York.

[3] Bjerve S. and Doksum, K. (1993) Correlation curves: measures of association as functions of covariates values. *Annals of Statistics* **21**, 890-902.

[4] Blyth, S. (1994) Local divergence and association. *Biometrika* **81**, 579–584.

[5] Block, H. W., Sampson A. R. and Savits, T. H. (1990) (eds) *Topics in statistical dependence*, IMS Lecture Notes-Monograph Series, **16** IMS, Hayward, CA.

[6] Borkar, V. S. (1993) White-noise representations in stochastic realization theory. *SIAM J. Control Optim.* **31** 1093–1102.

[7] Bradley, R. (2005) Basic properties of strong mixing conditions. A Survey and some open questions. *Probability Surveys* **2**, 107-144

[8] Bryk, A. and Mielniczuk, J. (2005) Asymptotic properties of kernel density estimates for linear processes: application of projection method. *Nonparametric Statistics* **14**, 121-133.

[9] Capéraà, P. and Genest, C. (1990). Concepts de dépendance et ordres stochastiques pour des lois bidimensionelles. *Canadian Journal of Statistics* **18**, 315-326.

[10] Cowell, R. G., Dawid, A. P. and Spiegelhalter, D. J. (1999) *Probabilistic Networks and Systems*, Springer, New York.

[11] Crutchfeld, J.P. and Feldman, D.P. (2003) Regularities unseen, randomness observed: The entropy convergence hierarchy. *Chaos* **15**, 25-54.

[12] Deák, I. (1990) *Random Numbers Generators and Simulation*, Akadémiai Kiadó, Budapest.

[13] Dedecker, J., Merlevède, F. (2002) Necessary and sufficient conditions for the conditional central limit theorem. *Ann. Probab.* **30**, 1044–1081.

[14] Dedecker, J., P. Doukhan, G. Lang, J.R. Leon R., S. Louhichi and C. Prieur (2007) *Weak Dependence: With Examples and Applications*, Springer, New York.

[15] Diaconis, P. and Freedman, D. (1999) Iterated random functions. *SIAM Review* **41** 41-76.

[16] Doksum, K. and Froda, S. M. (2000) Neighborhood correlation. *J. Statist. Plan. Inf.* **91**, 267–294.

[17] Doksum, K. and Samarov, A. (1995) Nonparametric estimation of global functionals and measure of the explanatory powers in regression. *Annals of Statistics* **23**, 1443-1473.

[18] Doukhan, P. and Louhichi, S. (1999) A new weak dependence condition and applications to moment inequalities. *Stochastic Process. Appl.* **84** 313–342.

[19] Drouet Mari, D. and Kotz, S. (2001) *Correlation and dependence*, Imperial College Press, London.

[20] Efron, B. and Stein, C. (1981) The jackknife estimate of variance. *Annals of Statistics* **9**, 586–596.

[21] Embrechts, P., McNeil, A. and Straumann, D. (2002) Correlatation and dependence in risk management: properties and pitfalls. In: *Risk Management: Value at Risk and Beyond.* (M.A.H. Dempster, ed.), Cambridge University Press, 176-223.

[22] Esary, J. D., Proschan, F. and Walkup, D.W. (1967) Association of random variables with applications, *Annals of Mathematical Statistics* **38**, 1466–1474.

[23] Esary, J. D. and Proschan, F. (1972) Relationships between some concepts of bivariate dependence. *Annals of Mathematical Statistics* **43**, 651-655.

[24] Glymour, C. (2006) Causation-II. Encyclopedia of Statistical Sciences, Wiley. 782–794

[25] Gordin, M.I. (1969) The central limit theorem for stationary processes. Dokl. Akad. Nauk SSSR 188, 739–741

[26] Gordin, M.I. and Lifsic, B.A.(1973) The central limit theorem for stationary Markov processes. *Dokl. Akad. Nauk SSSR* **239**, 392–393

[27] Granger, C.W.J. (1969) Investigating causal relations by econometric models and cross-spectral methods. *Econometrica* **37**, 424–438.

[28] Hannan, E.J. (1979) The central limit theorem for time series regression. *Stochastic Processes Appl.*, **9**, 281–289

[29] Joe, H. (1997) *Multivariate models and dependence concepts*, Chapman and Hall, London.

[30] Jogdeo, K. (1978) On a probability bound of Marshall and Olkin. *Ann. Statist.* **6**, 232–234.

[31] Jogdeo, K. (2006) Concepts of dependence. Encyclopedia of Statistical Sciences, Wiley, 163–1647.

[32] Kalikow, S. A. (1982) $T, T^{-1}$ transformation is not loosely Bernoulli. *Ann. Math.* **115**, 393–409.

[33] Kallianpur, G. (1981) Some ramifications of Wieners ideas on nonlinear prediction. In: *Norbert Wiener, Collected Works with Commentaries*, MIT Press, Mass., 402–424.

[34] Lai, C. D. and Xie, M. (2006) *Stochastic Ageing and Dependence for Reliability.* Springer, New York.

[35] Lehmann, E. L. (1966) Some concepts of dependence. *Ann. Math. Statist.* **37**, 1137–1153.

[36] Li, H. , Scarsini, M. and Shaked, M. (1996) Linkages: A tool for construction of multivariate distributions with nonoverlapping multivariate marginals. *J. Multivariate. Anal.* **56**, 20–41.

[37] Liggett, T.M. (1985) *Interacting Particle Systems*, Springer, New York

[38] Mellor, D. H. (1998) *The Facts of Causation*, Routledge, New York.

[39] Merlevède and F., Peligrad, M. (2006) On the weak invariance principle for stationary sequences under projective criteria. *J. Theor. Probab.* **19**, 647–689

[40] J. P. Morgan (1996) *RiskMetrics.* Technical Document. New York.

[41] Mosteller, F. and Tukey, J.W. (1977) *Data Analysis and Regression*, Addison–Wesley Reading, Mass.

[42] Newman, C.M. (1984) Asymptotic independence and limit theorems for positively and negatively dependent random variables. In: *Inequalities in Statistics and Probability* (Y.L. Tong, ed.), IMS Lecture Notes - Monograph Series 7, 127–140.

[43] Ornstein, D. S. (1973) An example of a Kolmogorov automorphism that is not a Bernoulli shift. *Advances in Math.* **10** 49–62

[44] Rényi, A. (1959) On measures of dependence. *Acta Math. Acad. Sci. Hungar.* **10** 441–451

[45] Rosenblatt, M. (1952). Remarks on a multivariate transformation. *Ann. Math. Statist.* **23** 470–472.

[46] Rosenblatt, M. (1959). Stationary processes as shifts of functions of independent random variables. *J. Math. Mech.* **8**, 665–681.

[47] Rosenblatt, M. (1971) *Markov Processes. Structure and Asymptotic Behavior*, Springer, New York.

[48] Rosenblatt, M. (2009) A comment on a conjecture of N. Wiener. *Statist. Probab. Letters,* **79**, 347–348

[49] Rubin, D. (1990) Formal modes of statistical inference for causal effects. *Journal of Statistical Planning and Inference* **25**, 279–292.

[50] Rüschendorf, L. (1981) Stochastically ordered distributions and monotonicity of the OC-function of sequential probability ratio tests. *Mathematische Operationsforschung und Statistik. Series Statistcis* **12**, 327–338.

[51] Rüschendorf, L. and de Valk, V.(1981) On regression representation of stochastic processes. *Stochastic Processes and Their Applications* **46**, 183–198.

[52] Scarsini, M. (1984) On measures of concordance. *Stochastica* **8**, 201–218.

[53] Skaug, H. J. and Tjøstheim, D. (1993) A nonparametric test of serial independence based on the empirical distribution function. *Biometrika* **80** 591–602.

[54] Stigler, S. M. (1986) *The history of statistics*, Harvard University Press, Cambridge, MA.

[55] Szekli, R. (1995) *Stochastic ordering and dependence in applied probability*, Springer, New York

[56] Tjøstheim, D. (1996) Measures of dependence and tests of independence. *Statistics* **28** 249–284.

[57] Tong, H. (1990) *Nonlinear time series analysis: A dynamic approach.*, Oxford University Press, Oxford.

[58] Volńy, D. (1993) Approximating martingales and the central limit theorem for strictly stationary processes. *Stochastic Processes and their Applications* **44**, 41–74

[59] Wiener, N. (1958) *Nonlinear Problems in Random Theory.*, MIT Press, Cambridge, MA.

[60] Woodroofe, M. (1992) A central limit theorem for functions of a Markov chain with applications to shifts. *Stochastic Processes and Their Applications* **41**, 33–44.

[61] Wu,W.B. and Woodroofe, M. (2004) Martingale approximations for sums of stationary processes. *Ann. Probab.* **32**, 1674–1690.

[62] Wu, W. B. (2005) Nonlinear system theory: Another look at dependence. *Proceedings of the National Academy of Sciences, USA*, **102**, 14150–14154.

[63] Wu, W. B. (2007) Strong invariance principles for dependent random variables. *Ann. Probab.*, **35**, 2294–2320.

[64] Wu, W. B. (2008) Empirical processes of stationary sequences. *Statistica Sinica*, **18** 313–333.

[65] Wu, W. B. and Shao, X. (2004) Limit theorems for iterated random functions. *Journal of Applied Probability*, **41** 425–436.

# Robust regression with infinite moving average errors

Patrick J. Farrell and Mohamedou Ould-Haye

**Abstract** This paper deals with the so-called S-estimation, which is known to have up to 50% breakdown point and therefore yielding one of the most robust estimators against outlier values, in a multiple linear regression when the errors are dependent. We mainly focus on the case when these errors exhibit a long range dependence pattern. We establish asymptotic normality of the slope estimator via some maximal inequality results à la Wu (2003) for dependent data, more suitable here than Kim and Pollard (1990) ones used by Davies (1990) in the i.i.d. case.

## 1 Introduction

In this paper we deal with the problem of estimating the slope in a multiple linear regression when the errors are assumed to exhibit a so-called long range dependence (LRD) or equivalently long memory pattern. This property means that the responses are so strongly dependent that their covariance function dies down very slowly, and therefore it is not summable. This pattern is in a sharp contrast with the i.i.d. case where the data have no memory at all. An intermediate case called short memory or short range dependence (SRD) has also been investigated by some authors. The slope estimator we consider here is the so-called S-estimator. We should mention first that to the best of our knowledge, the available literature on linear regression with LRD errors deals only with classical estimation such as least square estimation (LSE) and the like (see for example Robinson and Hidalgo (1997), Dahlhaus (1995)), or M-estimation and alike (see for example Giraitis, Koul, and Surgailis

P. Farrell
Carleton University, 1125 Colonel By Dr. Ottawa, Ontario, Canada, K1S 5B6,
e-mail: pfarrell@math.carleton.ca

M. Ould-Haye
Carleton University, 1125 Colonel By Dr. Ottawa, Ontario, Canada, K1S 5B6,
e-mail: ouldhaye@math.carleton.ca

P. Doukhan et al. (eds.), *Dependence in Probability and Statistics*,
Lecture Notes in Statistics 200, DOI 10.1007/978-3-642-14104-1_8,
© Springer-Verlag Berlin Heidelberg 2010

(1996)). S-estimation has been considered by some authors, but only for i.i.d errors. (See for example the seminal paper of Davies (1990)). The advantage of S-estimators relative to those based on LSE is mainly the fact that the former have the nice property of being highly robust to outlier values in terms of the breakdown point notion, which we will discuss in the next section. Specifically, we will consider the model

$$y_t = x_t(n)'\beta + \varepsilon_t, \qquad t = 1, \ldots, n, \qquad \text{where } ' \text{ is the transpose.} \qquad (1)$$

Here the design vector $x_t(n) \in \mathbb{R}^k$ is assumed to be deterministic, and the slope vector $\beta \in \mathbb{R}^k$ is unknown. The errors $\varepsilon_t$ are to possess the property of LRD, with an infinite moving average MA($\infty$) representation

$$\varepsilon_t = \sum_{j=1}^{\infty} a_j \xi_{t-j} \qquad (2)$$

with

$$a_j = j^{-\alpha} L(j), \qquad \alpha \in (1/2, 1), \qquad (3)$$

where $L$ is a slowly varying function at infinity (e.x. $L(j) = \ln(j)$), and $\xi_t$ are zero-mean i.i.d innovations. Note that condition 3) guarantees that both i) $a_j$ are square-summable, so the process $\varepsilon_t$ is well defined, and that ii) the covariance function

$$\gamma(n) := \text{Cov}(\varepsilon_t, \varepsilon_{t+n}) = n^{1-2\alpha} L^2(n)$$

is not summable, so the error-process $\varepsilon_t$ is of LRD.

## 2 S-estimators

Here we borrow some notation from Davies (1990). Let $\rho : \mathbb{R} \mapsto [0,1]$ be a symmetric differentiable function, non-increasing on the positive half real line with $\rho(0) = 1$, and with bounded support $[-e, e]$ (or simply constant beyond $|e|$).
Assume that $\varepsilon_t$ has marginal density $\sigma f$, where $\sigma \in (0, \infty)$ is an unknown scale parameter, and let

$$R(y, s) := \mathbb{E}\left[\rho\left(\frac{\varepsilon_1/\sigma - y}{s}\right)\right] = \int_{-\infty}^{\infty} \rho\left(\frac{u-y}{s}\right) f(u)du. \qquad (4)$$

Given $0 < \varepsilon < 1$ and a sample of observations $(x_1, y_1), \ldots, (x_n, y_n)$, the corresponding S-estimator of $(\beta, \sigma)$ is then the value $(\widehat{b}_n, \widehat{s}_n)$ of $(b, s) \in \mathbb{R} \times (0, \infty)$ that minimize $s$ subject to

$$\frac{1}{n}\sum_{t=1}^{n} \rho\left(\frac{y_t - x_t'b}{s}\right) \geq 1 - \varepsilon. \qquad (5)$$

The quantity $\varepsilon$ is used to choose the so-called theoretical breakdown point of the estimator. This breakdown point is a measure of the robustness of the estimator to outlier values and is defined by $\varepsilon^* = \min(\varepsilon, 1 - \varepsilon)$. Usually we choose a function $\rho$ that satisfies

$$R(0,1) = 1 - \varepsilon. \tag{6}$$

A typical example of such a function $\rho$ is

$$\rho(u) = \left(1 - \frac{u^2}{e^2}\right)^2, \qquad -e \leq u \leq e,$$

where $e$ is chosen to guarantee (6).

For a sample of observations $Z = ((x_i, y_i), i = 1, \ldots, n)$, the empirical breakdown point of an estimator $T(Z)$ of a parameter $\theta$ can be defined as follows: First consider all possible corrupted samples $Z'$ that are obtained by replacing any $m$ points of the original data by any arbitrary values (or bad outliers). Then the empirical breakdown point of $T$ associated with the sample $Z$ is given by

$$\varepsilon_n^*(T, Z) = \min_{1 \leq m \leq n} \left\{ \frac{m}{n}, \quad \|T(Z) - T(Z')\| = \infty \right\}.$$

In other words, the empirical breakdown point of the estimator $T$ is the smallest fraction of contaminations that cause the estimator $T$ to take on values arbitrarily far from $T(Z)$. For example, it is well known that the least square estimator (LSE) has the smallest empirical breakdown point $1/n$ which tends to zero as $n \to \infty$, leading to the fact that LSE has theoretical breakdown point $\varepsilon* = 0$, and making LSE extremely sensitive to outliers. On the other hand, the S-estimator has the largest breakdown point of up to 50%. Clearly beyond this level, it becomes impossible to distinguish between the "good" and "bad" parts of the sample. For more details on this notion of breakdown point, we refer the reader to Rousseeuw and Leroy's book (1987).

Before establishing the asymptotic properties of S-estimators in linear regression with LRD errors, we note that without restriction of generality, we can choose $(\beta, \sigma) = (0, 1)$. Actually, like the LSE, the S-estimator also has the desirable properties of being regression, scale, and affine equivariant. We recall quickly that for an estimator $T$, these notions correspond respectively, for any vector $v$, scalar $c$, and nonsingular matrix $A$, to (see Rousseeuw and Leroy 1987)

$$T(x_i, y_i + x_i v) = T(x_i, y_i) + v, \quad T(x_i, cy_i) = cT(x_i, y_i), \quad T(x_i A, y_i) = A^{-1} T(x_i, y_i).$$

## 3 S-estimators' Asymptotic Behavior

We make the following assumption on the design vector $x_t(n)$, to take into consideration the LRD pattern (3) of errors $\varepsilon_t$:

$$\sum_{i=1}^{n} x_i(n)x_i(n)' = nI_k, \tag{7}$$

where $I_k$ is the $k \times k$ identity matrix. This implies that

$$\sum_{i=1}^{n} \|x_i(n)\|^2 = kn. \tag{8}$$

First we mention that the estimators $\widehat{s}_n$ and $\widehat{b}_n$ are consistent, i.e. they converge in probability respectively to $\sigma$ and $\beta$. The proof is similar to Davies' (1990) and is based on the argument of strong law of large numbers. This argument still holds here. However, the difference is that in Davies (1990) errors $\varepsilon_t$ are i.i.d. Here although they are not i.i.d, they form an ergodic process (see Cramer and Leadbetter book (1967)) and therefore we can still apply the ergodicity theorem.

### 3.1 Weak Convergence of estimators

We will prove asymptotic normality of $\widehat{b}_n$ with rate $n^{\alpha-1/2}$. This convergence rate is much slower than the usual $\sqrt{n}$ and the difference is due to the slow decay of the covariance function of errors $\varepsilon_t$. The method we use is completely different from the one used by Davies (1990) which is based on i) a Pollard (1989) maximal inequality to establish tightness and ii) Lindeberg type conditions to obtain finite dimensional distributions convergence. Here we use some martingale difference sequence (MDS) methods for linear processes. As for $\widehat{s}_n$, it seems that, unlike in the i.i.d. case, it will not have systematically asymptotic normality when errors are LRD. We will not pursue this problem, although we will be showing that it is a $O_P(n^{1/2-\alpha})$

To establish asymptotic normality for the estimator $\widehat{b}_n$ we will use theorem 2.7 of Kim and Pollard(1990) which, for the sake of clarity, we reproduce here using our notation, and later will be referred to as KP theorem.

**Theorem 3.1. KP** *Let $Z_n(t)$ be a sequence of stochastic processes, $t \in \mathbb{R}^k$ such that*

*i)*

$$Z_n(t) \Longrightarrow Z(t) \quad in \ B_{loc}(\mathbb{R}^k),$$

*the space of all locally bounded functions, and $Z$ is concentrated on $C(\mathbb{R}^k)$, the subspace of continuous functions. Both spaces are equipped with uniform convergence on compact sets,*

*ii)*

$$argmax Z_n(t) = O_P(1), \quad i.e. \ this \ sequence \ is \ tight.$$

*iii) $Z(t)$ has a unique maximum.*

*Then we have as $n \to \infty$,*

$$argmax Z_n(t) \Longrightarrow argmax Z(t).$$

Before establishing the asymptotic normality for $\widehat{b}_n$, we prove the following for $\widehat{s}_n$ and $\widehat{b}_n$ in order to satisfy condition ii) of KP theorem.

**Proposition 0.1.** *We have*

$$n^{\alpha-1/2}\left(\widehat{s}_n - 1\right) = O_P(1). \tag{9}$$

$$n^{\alpha-1/2}\widehat{b}_n = O_P(1). \tag{10}$$

The proof of Proposition 1 is based on a result of Wu (2003), theorem 2, establishing a maximum inequality for LRD variables similar to Pollard's maximal inequality used in the i.i.d. case. Let's first present Wu's result here briefly, and explain how we should adapt it to our context of a function of several variables. We have

$$\mathbb{E}\left[\sup_{K \in \kappa}\left(\sum_{j=1}^{n}\left(K(\varepsilon_j) - K'_\infty(0)\right)\right)^2\right] = O\left(n^{3-2\alpha}\ell^2(n)\right), \tag{11}$$

for some nice class of functions $\kappa$. Here $K_\infty(x) = \mathbb{E}[K(\varepsilon_1 + x)]$. The class $\kappa$ can be considered here as the class of the functions indexed by $b, s$ satisfying $\|b\| \leq 1$, $|s - 1| < 1/2$, and its elements are given by

$$K(u) = \rho\left(\frac{u - b}{s}\right) - \rho(u).$$

We provide below an adaptation of theorem 2 of Wu (2003) to take into consideration $x_j$. Here elements of $\kappa$ will be functions of two parameters $\varepsilon_j$ and $x_j$. We will borrow some notation from Ho and Hsing (1996, 1997). Let $p$ be a nonnegative integer. Denote for $r = 0, \ldots, p$,

$$U_{nr} = \sum_{0 \leq j_1 < \ldots < j_r} a_{j_1}\xi_{n-j_1}\cdots a_{j_r}\xi_{n-j_r}.$$

Further, denote for a $p$-differentiable function $K$,

$$K_\infty(x, x_j) = \mathbb{E}\left[K(\varepsilon_1 + x, x_j)\right]$$

and

$$K_\infty^{(r)}(0, x_j) = \frac{\partial^r K_\infty}{\partial u^r}(u, x_j)\Big|_{u=0}.$$

Let

$$S_n(K, p) = \sum_{j=1}^{n}\left[K(\varepsilon_j, x_j) - \sum_{r=0}^{p}K_\infty^{(r)}(0, x_j)U_{j,r}\right]. \tag{12}$$

**Theorem 3.2.** *We have the following*

$$\mathbb{E}\left[\sup_{K \in \kappa}|S_n(K, p)|^2\right] = O\left(n\log^2 n\right) + E(n, p)$$

*where*

$$E(n,p) = O(n), \qquad O\big(n^{2-(p+1)(2\alpha-1)}\big), \quad \text{or } O(n)(\log n)^2$$

*according to*

$$(p+1)(2\alpha-1) > 1, \qquad (p+1)(2\alpha-1) < 1, \qquad \text{or } (p+1)(2\alpha-1) = 1.$$

*Proof.* Similar to Wu (2003), the key component in the proof is to write the centered partial sums in (12) in terms of the empirical process

$$S_n(K,p) = -\int_{\mathbb{R}} g_K(t)S_n(t,p)dt \tag{13}$$

where

$$S_n(t,p) = \sum_{j=1}^{n}\left(1_{\varepsilon_j \le t} - \sum_{r=0}^{p} F^{(r)}(0)U_{j,r}\right) := \sum_{j=1}^{n} L(\widetilde{\varepsilon}_j, t)$$

where $\widetilde{\varepsilon}_j = (\ldots, \varepsilon_{j-2}, \varepsilon_{j-1}, \varepsilon_j)$.

For the sake of simplicity, we present here the adaptation where $x_j$ are real numbers rather than vectors. Assume that

$$K(\varepsilon, x) = \int_0^{\varepsilon} g_K(t,x)dt$$

and that

$$g_K(t,x) = \int_0^{x} h_K(t,u)du.$$

Successively, as in the proof of Theorem 1 of Wu (2003), we obtain for fixed $x_j$ and $\varepsilon_j$

$$K(\varepsilon_j, x_j) - \sum_{r=0}^{p} K_{\infty}^{(r)}(0, x_j)U_{j,r} = -\int_{\mathbb{R}} g_K(t, x_j)\left(1_{\varepsilon_j \le t} - \sum_{r=0}^{p} F^{(r)}(0)U_{j,r}\right)$$

$$= -\int_{\mathbb{R}} g_K(t, x_j)L(\widetilde{\varepsilon}_j, t)dt$$

$$= -\int_{\mathbb{R}} \int_a^b \left[1_{u \le x_j} h_K(t,u)L(\widetilde{\varepsilon}_j, t)\right]dudt$$

for some $a, b$, and therefore putting

$$S_n(t,u,p) = \sum_{\substack{j=1 \\ u \le x_j}}^{n} L(\widetilde{\varepsilon}_j, t),$$

we obtain the counterpart of (13) in the form

$$S_n(K,p) = -\int_{\mathbb{R}} \int_a^b h_K(t,u)S_n(t,u,p)dudt \tag{14}$$

Let for $\gamma > 0$, $w_\gamma(t) = (1 + |t|)^\gamma$, and let $\kappa(\gamma)$ be the class of functions $K$ such that

$$\sup_{K \in \kappa(\gamma)} \int_{\mathbb{R}} \int_a^b h_K^2(t, u) w_{-\gamma}(dt) du \leq 1.$$

Then we can write using Cauchy's inequality, and (14)

$$\mathbb{E}\left[\sup_{K \in \kappa(\gamma)} |S_n(K, p)|^2\right]$$

$$\leq \mathbb{E}\left[\sup_{K \in \kappa(\gamma)} \int_{\mathbb{R}} \int_a^b h_K^2(t, u) w_{-\gamma}(dt) du \int_{\mathbb{R}} \int_a^b |S_n(t, u, p)|^2 w_\gamma(dt) du\right]$$

$$\leq \int_{\mathbb{R}} \int_a^b \|S_n(t, u, p)\|^2 w_\gamma(t) dt du$$

where $\|.\|$ is the $L^2$ norm.

For any integer $j$, let $\mathscr{P}_j$ be the projection on the subspace generated by $(\ldots, \xi_{j-2}, \xi_{j-1}, \xi_j)$. Similarly to the proof of Lemma 11 of Wu(2003), we have that $\mathscr{P}_j S_n(t, u, p)$ are orthogonal for $j \leq n$ and $\mathscr{P}_j L(\widetilde{\varepsilon}_\ell, t) = 0$ for $\ell < j$. Hence by the Parseval equality we have that for every fixed $u$

$$\|S_n(t, u, p)\|^2 = \sum_{j=-\infty}^n \|\mathscr{P}_j S_n(t, u, p)\|^2 = \sum_{j=-\infty}^n \left(\sum_{\ell=\max(1,j)}^n (\mathscr{P}_j L(\widetilde{\varepsilon}_\ell, t, u))^2\right)$$

$$\leq \sum_{j=-\infty}^n \left(\sum_{\ell=\max(1,j)}^n (\mathscr{P}_j L(\widetilde{\varepsilon}_\ell, t, p))^2\right) = \|S_n(t, p)\|^2$$

and hence the result of Lemma 11 of Wu (2003) still holds for $S_n(t, u, p)$, i.e.

$$\int_{\mathbb{R}} \int_a^b \|S_n(t, u, p)\|^2 \omega(dt) du \leq (b - a) \int_{\mathbb{R}} \|S_n(t, p)\|^2 \omega(dt) = O(n \log^2 n) + E(n, p).$$

Taking $p = 1$ in Theorem 4 we obtain the following weak reduction principle for all $1/2 < \alpha < 1$,

$$n^{(2\alpha-1)-2} \mathbb{E}\left[\sup_{K \in \kappa(\gamma)} |S_n(K, 1)|^2\right] = o(1).$$

Therefore we have

$$n^{\alpha-3/2}\left[\sum_{j=1}^n \left(K(\varepsilon_j, x_j) - K_\infty(0, x_j)\right)\right] = n^{\alpha-3/2} \sum_{j=1}^n K_\infty'(0, x_j)\varepsilon_j + o_P(1), \quad (15)$$

uniformly in $K \in \kappa$. $\diamond$

## 4 Proof of Proposition 1

**Proof of (9):** In what follows we assume $\rho$ is differentiable and $\rho' = \psi$. Let

$$A(\rho) := -\int_{-\infty}^{\infty} u\psi(u)f(u)du = \frac{\partial R(0,s)}{\partial s}R(0,s)\Big|_{s=1},$$

We consider the set of functions

$$\kappa = \left\{ K(u,v) = \rho\left(\frac{u-vb}{s}\right) - \rho\left(\frac{u}{s}\right), \quad b \in \mathbb{R}^k, \|b\| < \delta, \quad s \in \mathbb{R}, \quad |s| < 2 \right\}.$$

It is easy to verify that $\kappa$ satisfies the conditions of Wu (2003). Using (15) we obtain that

$$Z_n(b,s) := n^{\alpha-3/2} \sum_{j=1}^n \left( \rho\left(\frac{\varepsilon_j - x'_j b}{s}\right) - \rho\left(\frac{\varepsilon_j}{s}\right) - \left(R(x'_j b, s) - R(0,s)\right) \right)$$

$$= n^{\alpha-3/2} \sum_{j=1}^n \varepsilon_j \frac{\partial}{\partial x}\left(R(x'_j b - x, s) - R(-x, s)\right)\Bigg|_{x=0} + o_P(1).$$

Assuming that $R(u,v)$ is twice differentiable with respect to $u$, we get that

$$n^{\alpha-3/2} \sum_{j=1}^n \varepsilon_j \frac{\partial}{\partial x}\left(R(x'_j b - x, s) - R(-x, s)\right)\Bigg|_{x=0}$$

$$= Dn^{\alpha-3/2} \sum_{j=1}^n \varepsilon_j x'_j b + o_P(1),$$

for some constant $D$. We know from Giraitis et al. (1996) that last sequence converges in distribution to $b'Z$ where $Z$ is a normal distribution. Hence as $b$ goes to zero, this sequence will converge to zero in probability. Therefore we have

$$\lim_{\delta \to 0}\lim_{n \to \infty}\left(\sup_{\|b\|<\delta, |s|<2} |Z_n(b,s)|\right) = 0 \qquad \text{in probability.} \tag{16}$$

Since $\widehat{b}_n \to 0$ and $\widehat{s}_n \to 1$ in probability, we get

$$\limsup_{n \to \infty} P\left(|Z_n(\widehat{b}_n, \widehat{s}_n)| > \eta\right) \leq \limsup_{n \to \infty} P\left(|Z_n(\widehat{b}_n, \widehat{s}_n)| > \eta, \quad \|\widehat{b}_n\| \leq \delta, \quad |\widehat{s}_n| < 2\right)$$

$$+ \limsup_{n \to \infty} P(\|\widehat{b}_n\| > \delta) + \limsup_{n \to \infty} P(|\widehat{s}_n| > 2)$$

$$= \limsup_{n \to \infty} P\left(|Z_n(\widehat{b}_n, \widehat{s}_n)| > \eta, \quad \|\widehat{b}_n\| \leq \delta, \quad |\widehat{s}_n| < 2\right)$$

$$\to 0 \text{ as } \delta \to 0$$

by (16). Using (5), we then obtain

$$
1 - \varepsilon \leq \frac{1}{n} \sum_{j=1}^{n} \left( \rho \left( \frac{\varepsilon_j - x_j' \widehat{b}_n}{\widehat{s}_n} \right) \right)
$$

$$
= \frac{1}{n} \sum_{j=1}^{n} \rho \left( \frac{\varepsilon_j}{\widehat{s}_n} \right) + \frac{1}{n} \sum_{j=1}^{n} \left( R(x_j' \widehat{b}_n, \widehat{s}_n) - R(0, \widehat{s}_n) \right) + o_P \left( n^{1/2 - \alpha} \right)
$$

$$
\leq \frac{1}{n} \sum_{j=1}^{n} \rho \left( \frac{\varepsilon_j}{\widehat{s}_n} \right) + o_P \left( n^{1/2 - \alpha} \right) \tag{17}
$$

Similarly if we apply Theorem 2 of Wu (2003) to the following normalized partial sum

$$
Z_n'(s) = n^{\alpha - 3/2} \sum_{j=1}^{n} \left[ \rho \left( \frac{\varepsilon_j}{s} \right) - R(0, s) - \left[ \rho(\varepsilon_j) - R(0, 1) \right] \right],
$$

we get that

$$
n^{\alpha - 3/2} \sum_{j=1}^{n} \left( \rho \left( \frac{\varepsilon_j}{s} \right) - R(0, s) \right) = n^{\alpha - 3/2} \sum_{j=1}^{n} \left( \rho(\varepsilon_j) - R(0, 1) \right) + o_P(1). \tag{18}
$$

Hence, multiplying by $n^{1/2 - \alpha}$, and using (17) we get

$$
R(0, \widehat{s}_n) - R(0, 1) = \frac{1}{n} \sum_{j=1}^{n} \rho \left( \frac{\varepsilon_j}{\widehat{s}_n} \right) - \frac{1}{n} \sum_{j=1}^{n} \rho(\varepsilon_j) + o_P \left( n^{1/2 - \alpha} \right)
$$

$$
\geq - \frac{1}{n} \sum_{j=1}^{n} \left( \rho(\varepsilon_j) - (1 - \varepsilon) \right) + o_P \left( n^{1/2 - \alpha} \right)
$$

Now using the fact that $R(0, s)$ is differentiable at $s = 1$ and has derivative $A(\rho)$, we obtain as $s \to 1$,

$$
R(0, s) - R(0, 1) = A(\rho)(s - 1) + o(s - 1)
$$

and therefore

$$
A(\rho)(\widehat{s}_n - 1) \geq - \frac{1}{n} \sum_{j=1}^{n} \left( \rho(\varepsilon_j) - (1 - \varepsilon) \right) + o_P \left( n^{1/2 - \alpha} \right). \tag{19}
$$

Note that to achieve (19) we used the fact that if $U_n + o_P(U_n) \geq V_n + o_P(Z_n)$ then $U_n \geq V_n + o_P(Z_n)$.

Next, we can proceed as in Davies (1996), to obtain from (18) and the differentiability of $R(0, s)$ at $s = 1$, that

$$
A(\rho)(\widehat{s}_n - 1) \leq - \frac{1}{n} \sum_{j=1}^{n} \left( \rho(\varepsilon_j) - (1 - \varepsilon) \right) + o_P \left( n^{1/2 - \alpha} \right). \tag{20}
$$

Since $\rho$ is even, then the empirical mean

$$\frac{1}{n}\sum_{j=1}^{n}\left(\rho(\varepsilon_j)-(1-\varepsilon)\right)$$

multiplied by $n^{1-2\alpha}$ will converge weakly to a Rosenblatt distribution (see Ho and Hsing 1997, for example). Combining this with (19) and (20), we reach the conclusion that

$$\widehat{s}_n - 1 = O_P(n^{\alpha-1/2})$$

which completes the proof of (9) of Proposition 1.

**Proof of (10):** We assume the following two conditions of Davies (1990): $R(u,v)$ is twice differentiable with bounded second derivative at $u$. Also we assume that

$$\lim_{(u,v)\to(0,1)}\frac{R(u,v)-R(0,1)}{u^2} = -B(\rho) < 0. \tag{21}$$

For some $\eta_1, \eta_2$, and $n_0$, we have (with $I_k$ the $k \times k$ identity matrix),

$$\sum_{j=1}^{n}x_jx_j'\mathbf{1}_{\|x_j\|<\eta_1} - n\eta_2 I_k \text{ is positive definite for } n \geq n_0,$$

from which we can conclude that for all $b$, we have

$$\sum_{\substack{j=1 \\ \|x_j\|<\eta_1}}^{n}b'x_jx_j'b \geq n\eta_2\|b\|^2. \tag{22}$$

We will show an analogous result to Lemma 4.1 of Kim and Pollard (1990). If $b$ is bounded, then for every $\eta > 0$, there exists a random variable $M_n = O_P(1)$ such that we have

$$\frac{1}{n}\left|\sum_{j=1}^{n}x_j'\varepsilon_j b\right| \leq \eta|b|^2 + n^{1-2\alpha}M_n. \tag{23}$$

**Proof of (23)** For fixed $\omega$ let $M_n(\omega)$ be the infimum of those values of $M_n$ for which (23) holds. Then with

$$A(n,j) = \{b \text{ such that } (j-1)n^{1/2-\alpha} \leq b < jn^{1/2-\alpha}\},$$

we can write

$$P(M_n > m) \leq P\left[\exists b : \frac{1}{n}\left|\sum_{j=1}^{n} x'_j \varepsilon_j b\right| \geq \eta |b|^2 + n^{1-2\alpha} m^2\right]$$

$$\leq P\left[\exists b \in A(n,j) : \ n^{2\alpha-1}\frac{1}{n}\left|\sum_{j=1}^{n} x'_j \varepsilon_j b\right| \geq \eta(j-1)^2 + m^2\right]$$

$$\leq \sum_{j=1}^{\infty} P\left[\sup_{b \in A(n,j)} \frac{n^{\alpha-1/2}}{n}\left(n^{\alpha-1/2}\left|\sum_{j=1}^{n} x'_j \varepsilon_j b\right|\right) \geq \eta(j-1)^2 + m^2\right]$$

$$\leq n^{2\alpha-1} \sum_{j=1}^{\infty} \sup_{b \in A(n,j)} \frac{b'X'R_n Xb}{\left((\eta(j-1)^2 + m^2\right)^2} \quad \text{by Markov's inequality}$$

$$\leq C \sum_{j=1}^{\infty} \frac{j^2}{\left(\eta(j-1)^2 + m^2\right)^2} \quad \text{where } R_n \text{ is the } \varepsilon_j \text{ covariance matrix.}$$

This sum goes to zero as $m \to \infty$, uniformly in $n$. Hence $M_n$ is tight and therefore $O_P(1)$.

Next we apply Theorem 4 with $p = 1$ and for fixed $b, s$,

$$K(\varepsilon_j, x_j) = \rho\left(\frac{\varepsilon_j - x'_j b}{s}\right) - \rho\left(\frac{\varepsilon_j}{s}\right).$$

We obtain

$$\mathbb{E}\left[\left(\sup_{\|b\| \leq 1, |s-1| \leq 1/2} \sum_{j=1}^{n}\left\{\rho\left(\frac{\varepsilon_j - x'_j b}{s}\right) - \rho\left(\frac{\varepsilon_j}{s}\right) - \left(R(x'_j b, s) - R(0, s)\right)\right.\right.\right.$$
$$\left.\left.\left. - \varepsilon_j \frac{\partial}{\partial x}\left(R(x + x'_j b, s) - R(x, s)\right)\Big|_{x=0}\right\}\right)^2\right] = O(n^{4-4\alpha})$$

Then using the fact that

$$\frac{\partial}{\partial x}\left(R(x + x'_j b, s) - R(x, s)\right)\Big|_{x=0} = \frac{R''(u, s)}{s^2} x'_j b$$

for some $u$ between $0$ and $x'_j b$, that $R''$ is bounded, and the fact that $\hat{s}_n = 1 + O_P(n^{1/2-\alpha})$, we can then write

$$\frac{n^{2\alpha-1}}{n} \sum_{j=1}^{n}\left\{\rho\left(\frac{\varepsilon_j - x'_j b}{s}\right) - \rho\left(\frac{\varepsilon_j}{s}\right) - \left(R(x'_j b, s) - R(0, s)\right)\right\}$$
$$= \frac{n^{2\alpha-1}}{n} \sum_{j=1}^{n} x'_j b \varepsilon_j + O_P(1)$$

uniformly in $|b| \leq 1$ ad $|s-1| \leq 1/2$, and therefore we get from (23)

$$\frac{1}{n}\left|\sum_{j=1}^{n}\left\{\rho\left(\frac{\varepsilon_j-x_j'b}{s}\right)-\rho\left(\frac{\varepsilon_j}{s}\right)-\left(R(x_j'b,s)-R(0,s)\right)\right\}\right|\leq\eta|b|+n^{1-2\alpha}M_n$$

$$(24)$$

Now returning to the proof of (10). We note first that $\widehat{b}_n$ maximizes

$$\frac{1}{n}\sum_{j=1}^{n}\rho\left(\frac{\varepsilon_j-x_j'b}{\widehat{s}_n}\right)$$

and therefore we can write using (24)

$$0\leq\frac{1}{n}\sum_{j=1}^{n}\left\{\rho\left(\frac{\varepsilon_j-x_j'\widehat{b}_n}{\widehat{s}_n}\right)-\rho\left(\frac{\varepsilon_j}{\widehat{s}_n}\right)\right\}$$

$$\leq\eta\|\widehat{b}_n\|^2+n^{1-2\alpha}M_n^2+\frac{1}{n}\sum_{j=1}^{n}\left(R(x_j'\widehat{b}_n,\widehat{s}_n)-R(0,\widehat{s}_n)\right)$$

$$\leq\eta\|\widehat{b}_n\|^2+n^{1-2\alpha}M_n^2+\frac{1}{n}\sum_{\substack{j=1\\\|x_j\|<\eta_1}}^{n}\left(R(x_j'\widehat{b}_n,\widehat{s}_n)-R(0,\widehat{s}_n)\right)$$

$$=\eta\|\widehat{b}_n\|^2+n^{1-2\alpha}M_n^2+\frac{1}{n}\sum_{\substack{j=1\\\|x_j\|<\eta_1}}^{n}\left(-B(\rho)\widehat{b}_n'x_jx_j'\widehat{b}_n+o_P(\|\widehat{b}_n\|^2)\right)\text{ by (21)}$$

$$\leq\eta\|\widehat{b}_n\|^2+n^{1-2\alpha}M_n^2-\eta_2B(\rho)\|\widehat{b}_n\|^2+o_P(\|\widehat{b}_n\|^2)\text{ by (22)}$$

Choosing $\eta<\eta_2 B(\rho)$, we get

$$(\eta_2 B(\rho)-\eta)\|\widehat{b}_n\|^2\leq O_P(n^{1-2\alpha})$$

and hence $\widehat{b}_n=O_P(n^{1/2-\alpha})$ which proves the (10) of Proposition 1.
We end this section with the main result: we assume that the design matrix $X_n=(x_jx_j')_{j=1,\ldots,n}$ satisfies the conditions (G.1)-(G.3) and condition 6 with $m=k$ of Yajima (1991). We will also use the condition (8) stated earlier, and we assume in addition that

$$\lim_{(u,v)\to(0,1)}\frac{R'(u,v)-R'(0,1)}{u}=-C(\rho)\neq0.\qquad(25)$$

**Theorem 4.1.** *Let* $1/2<\alpha<3/4$. *I.e. we consider the very long memory case. Then*

$$n^{\alpha-1/2}\widehat{b}_n\Longrightarrow\mathcal{N}(0,Q)$$

*where*

$$Q=\frac{C^2(\rho)}{4B^2(\rho)}\lim_{n\to\infty}n^{2\alpha-3}X_n'R_nX_n.\qquad(26)$$

*Proof.* Again, applying Theorem 4 with $p=2$, we get uniformly in bounded $b$

$$n^{2\alpha-2} \sum_{j=1}^{n} \left\{ \rho\left(\frac{\varepsilon_j - x_j'b}{\widehat{s}_n}\right) - \rho\left(\frac{\varepsilon_j}{\widehat{s}_n}\right) \right\} = n^{2\alpha-2} \sum_{j=1}^{n} \left( R(x_j'b, \widehat{s}_n) - R(0, \widehat{s}_n) \right)$$

$$+ n^{2\alpha-2} \sum_{j=1}^{n} \varepsilon_j \frac{\partial}{\partial x} \left( R(x + x_j'b, \widehat{s}_n) - R(x, \widehat{s}_n) \right)\Bigg|_{x=0} \tag{27}$$

$$+ n^{2\alpha-2} \sum_{j=1}^{n} U_{j2} \frac{\partial^2}{\partial x^2} \left( R(x + x_j'b, \widehat{s}_n) - R(x, \widehat{s}_n) \right)\Bigg|_{x=0} + o_P(1)$$

Replacing $b$ by $bn^{\alpha-1/2}$ in (27) and using

$$\lim_{s \to 1} \frac{\partial}{\partial x} \left( R(x + x_j'b, s) - R(x, s) \right)\Bigg|_{x=0} = x_j'bC(\rho).$$

and

$$\lim_{s \to 1} \frac{\partial^2}{\partial x^2} \left( R(x + x_j'b, s) - R(x, s) \right)\Bigg|_{x=0} = x_j'bD(\rho)$$

for some coefficient $D(\rho)$, we obtain that as $n \to \infty$

$$n^{2\alpha-2} \sum_{j=1}^{n} \left\{ \rho\left(\frac{\varepsilon_j - x_j'bn^{1/2-\alpha}}{1 + \widehat{s}_n n^{1/2-\alpha}}\right) - \rho\left(\frac{\varepsilon_j}{1 + \widehat{s}_n n^{1/2-\alpha}}\right) \right\} \tag{28}$$

and

$$n^{2\alpha-2} \left( R(x_j'bn^{1/2-\alpha}, 1 + \widehat{s}_n n^{1/2-\alpha}) - R(0, 1 + \widehat{s}_n n^{1/2-\alpha}) \right)$$

$$+ C(\rho)n^{\alpha-3/2} \sum_{j=1}^{n} x_j'b\varepsilon_j \tag{29}$$

have the same limiting distributions. However by Giraitis et al. (1996) the later sum in (29) has limiting distribution $b'Z$ where $Z$ has multivariate normal distribution with mean zero and covariance matrix $Q$ in (26). By Yajima (1991) $Q$ is finite and different from zero.

Next, for the deterministic sum in (29), it can be evaluated similarly to Davies (1990) as equivalent to

$$-n^{2\alpha-2} \sum_{j=1}^{n} |x_jb|^2 n^{1-2\alpha} B(\rho) = -B(\rho)\|b\|^2$$

since

$$\sum_{j=1}^{n} x_j'x_j = nI_k.$$

Therefore the sequence of processes in (28) converges to $b'Z - B(\rho)\|b\|^2$ which has argmax equals $Z/(2B(\rho))$. Therefore by KP theorem the argmax of (28) which is

$n^{\alpha-1/2}\widehat{b}_n$ will converge in distribution to this limit. Of course we already showed that $n^{\alpha-1/2}\widehat{b}_n$ satisfies ii) of KP theorem. Note that unlike the i.i.d. case, here the limiting components are dependent. This may be due to the strong dependence of the initial variables. $\diamondsuit$

## 5 Discussion

In this paper, we have considered the problem of estimating the slope in a multiple linear regression when the error terms are assumed to exhibit a long memory pattern. Future research regarding the proposed estimator will include a simulation study to assess its properties for various sample sizes and different patterns of long range dependence in the errors.

We believe that the results obtained here can also be extended to the case of short range dependence, where the coefficients $a_j$ are summable, which is equivalent to the summability of covariances. This likely can be accomplished, for example, via some maximal inequalities similar to Kim and Pollard (1990), but for short range dependence. As in the i.i.d. case, the convergence rate would then be $\sqrt{n}$ . The results developed here can likely be extended to subordinated Gaussian processes as well, in particular for the situation when the Hermit rank is even. However, for such a situation, the limiting distribution would no longer be normal. A uniform weak approximation over a class of functions for Gaussian processes analogous to the one presented in this paper for linear ones would be required.

**Acknowledgements** The authors are grateful to the referee for valuable comments and suggestions. This work was supported through funds from the Natural Sciences and Engineering Research Council of Canada.

## References

[1] Cramer, H., Leadbetter, M.R. (1967) *Stationary and related stochastic processes*, New York: Wiley.
[2] Dahlhaus, R. (1995). Efficient location and regression estimation for long range dependent regression model. *Annals of stat.* **23**(3), 1029-1047.
[3] Davies, L. (1990) The asymptotics of S-estimators in the linear regression model, *Ann. Statist.* **18**(4), 1651-1675.
[4] Giraitis, L., Koul, H., Surgailis, D. (1996) Asymptotic normality of regression estimators with long memory errors, *Statistics and Probability Letters* **29** 317-335.
[5] Ho, H.-C, Hsing, T. (1996) On the asymptotic expansion of the empirical process of long memory moving average, *Ann. Statist.***24**, 992-1024.

[6] Ho, H.-C, Hsing, T. (1997) Limit theorems for functionals of moving averages *Ann. Statist.* **25**(4), 1636-1669.

[7] Kim, J., Pollard, D. (1990) Cube roots asymptotics, *Ann. Statist.* **18** 191-219.

[8] Pollard, D. (1989). Asymptotics via empirical processes. *Statistical science.* **4**(4), 341-354.

[9] Robinson, P.M., Hidalgo, J.F. (1997) Time Series regression with long range dependence, *Annals of Statistics* **25** 77-104.

[10] Rousseeuw, P.J., Leroy, A.M. (1987) *Robust regression and outlier detection* Wiley-InterScience, New York.

[11] Wu, W. B. (2003) Empirical processes of long memory sequences. *Bernoulli* **9**(5), 809-831.

[12] Yajima, Y. (1991) Asymptotic properties of LSE in a linear regression model with long memory stationary errors. *Ann. Statist.* **19**, 158-177.

# A note on the monitoring of changes in linear models with dependent errors

Alexander Schmitz and Josef G. Steinebach

**Abstract** Horváth et al. [8] developed a monitoring procedure for detecting a change in the parameters of a linear regression model having independent and identically distributed errors. We extend these results to allow for dependent errors, which need not be independent of the stochastic regressors, and we also provide a class of consistent variance estimators. Our results cover strongly mixing errors as an important example. Applications to autoregressive time series and near-epoch dependent regressors are discussed, too.

## 1 Introduction

In testing time series data for structural stability two different approaches can be chosen. There are retrospective procedures which deal with the detection of a structural break within an observed data set of fixed size, whereas sequential procedures check the stability hypothesis each time a new observation is available. Chu et al. [5] pointed out that the repeated application of a retrospective procedure each time new data arrive may result in a procedure that rejects a true null hypothesis of no change with probability one as the number of applications grows. Therefore they derived an alternative (sequential) testing procedure for detecting a change in the parameters of a linear regression model, after a stable training period of size $m$ (say). Their testing procedure is based on the first excess time of a detector over a boundary function, where the detector is a cumulative sum (CUSUM) type statistic of the residuals. The boundary function can be suitably chosen such that the test attains a prescribed asymptotic size $\alpha$ (say) and asymptotic power one as $m$ tends to infinity. Horváth et al. [8] extended these results and developed a CUSUM monitoring procedure for

Alexander Schmitz · Josef G. Steinebach
Mathematisches Institut, Universität zu Köln
Weyertal 86-90, 50931 Köln, Germany.
e-mail: schmitza@math.uni-koeln.de, jost@math.uni-koeln.de

P. Doukhan et al. (eds.), *Dependence in Probability and Statistics*,
Lecture Notes in Statistics 200, DOI 10.1007/978-3-642-14104-1_9,
© Springer-Verlag Berlin Heidelberg 2010

detecting an "early change", i.e. a change expected shortly after the monitoring has begun. Since they modeled the errors of the linear regression to be independent and identically distributed, [2] extended this CUSUM monitoring procedure further in order to obtain the right framework for monitoring changes in econometric data. To this end, they developed a testing procedure for monitoring a linear regression model with conditionally heteroskedastic errors.

Recently, Perron and Qu [17] introduced a retrospective multiple change-point analysis of a multivariate regression. They assumed strongly mixing errors, which need not necessarily be independent of the stochastic regressors. In this note we show that their dependence conditions permit the application of the CUSUM monitoring procedure as well. We explicitly describe the level of asymptotic independence which suffices to apply the monitoring.

The paper is organized as follows. In the next section, we specify the monitoring procedure and state the set of general conditions needed for its application. In Section 3 some examples are discussed. The first example gives precise conditions for the monitoring of a linear model with independent and identically distributed (i.i.d) errors and near-epoch dependent regressors. In the second example we show that our setup allows for an asymptotic $M$-dependence which, e.g., holds for conditionally heteroskedastic time series models. The third example illustrates the application of the monitoring procedure for testing parameter stability in an autoregressive (AR) model under the strong mixing condition. The proofs of our main results are given in Section 4.

## 2 The testing procedure

Throughout this paper we assume that all random variables are defined on a common probability space $(\Omega, \mathscr{F}, P)$. Our aim is to show that the sequential monitoring procedure for the linear model

$$y_i = \mathbf{x}_i^T \boldsymbol{\beta}_i + \varepsilon_i = \beta_{1,i} + x_{2,i}\beta_{2,i} + \cdots + x_{p,i}\beta_{p,i} + \varepsilon_i, \quad 1 \leq i < \infty, \qquad (1)$$

which was discussed in [8], continues to hold under an asymptotic independence relation specified below. We denote the $p \times 1$ random regressors by

$$\mathbf{x}_i^T = (1, x_{2,i}, ..., x_{p,i}), \qquad (2)$$

and set $\boldsymbol{\beta}_i^T = (\beta_{1,i}, ... \beta_{p,i})$ as the $p \times 1$ parameter vectors. We assume the parameter vectors to be constant over a training period of length $m$, i.e.

$$\boldsymbol{\beta}_i = \boldsymbol{\beta}_0, \quad 1 \leq i \leq m. \qquad (3)$$

This period is used as a reference for comparisons with future observations, noticing that the historical period and the future observation are asymptotically independent in a certain sense.

The monitoring procedure leads to a decision between the "no change" null hypothesis and the "change at unknown time $k^*$" alternative, i.e.

$$H_0 : \boldsymbol{\beta}_{m+i} = \boldsymbol{\beta}_0 \quad \text{for all } i \geq 1 \quad \text{versus}$$

$$H_A : \boldsymbol{\beta}_{m+k^*+i} = \boldsymbol{\beta}_* \neq \boldsymbol{\beta}_0 \quad \text{for some } 1 \leq k^* < \infty \text{ and for all } i \geq 0.$$

The parameter $k^*$ is called the change-point which is assumed to be unknown as well as the values of the parameters $\boldsymbol{\beta}_0$ and $\boldsymbol{\beta}_*$.

The monitoring procedure is defined via a stopping rule based on the first excess time $\tau_m$ of a detector $\widehat{Q}_m(\cdot)$ over a boundary function $g_m(\cdot)$. We choose

$$\tau_m = \inf \left\{ k \geq 1 : |\widehat{Q}_m(k)| > g_m(k) \right\}, \quad \text{where} \quad \inf \emptyset = \infty.$$

Following [8] the detector $\widehat{Q}_m(k) = \sum_{i=m+1}^{m+k} \widehat{\varepsilon}_i$ is a CUSUM type statistic of the residuals $\widehat{\varepsilon}_i = y_i - \mathbf{x}_i^T \widehat{\boldsymbol{\beta}}_m$, where the unknown regression parameter $\boldsymbol{\beta}_0$ is estimated by the least squares estimator $\widehat{\boldsymbol{\beta}}_m = \left( \sum_{i=1}^m \mathbf{x}_i \mathbf{x}_i^T \right)^{-1} \sum_{j=1}^m \mathbf{x}_j y_j$. We point out that the latter estimation relies only on the observations from the training period. The boundary function is chosen as $g_m(\cdot) = \sigma c g_m^*(\cdot)$. The parameter $\sigma$ is a normalizing constant which will be specified below. Therefore the only way to obtain a controlled asymptotic level $\alpha$ is to fix the critical constant $c = c(\alpha)$ such that, under the null hypothesis $H_0$,

$$\lim_{m \to \infty} P(\tau_m < \infty) = \alpha.$$

The main goal of this note is to derive a limiting distribution under a general asymptotic independence relation. It turns out that, if the mean regressor is not orthogonal to the parameter shift, the test has asymptotic power one.

We set $S_0(k) = \sum_{i=1}^k \varepsilon_i$, $S_m(k) = \sum_{i=m+1}^{m+k} \varepsilon_i$, and assume that the following conditions are satisfied: For each $m$, we can find two standard Wiener processes $\{W_{0,m}(t), 0 \leq t < \infty\}$ and $\{W_{1,m}(t), 0 \leq t < \infty\}$ and positive constants $\sigma$ and $\delta$ such that we have a uniform weak invariance principle (IP) over the training period, i.e.

$$\sup_{1 \leq k \leq m} k^{-1/(2+\delta)} |S_0(k) - \sigma W_{0,m}(k)| = O_P(1) \quad (m \to \infty), \tag{4}$$

together with a uniform weak IP for the monitoring sequence, i.e.

$$\sup_{1 \leq k < \infty} k^{-1/(2+\delta)} |S_m(k) - \sigma W_{1,m}(k)| = O_P(1) \quad (m \to \infty). \tag{5}$$

The next assumption describes the asymptotic independence relation. For each $n$ let $0 < t_1 < \cdots < t_n$ be consecutive points on the real line such that, for all $(x, \mathbf{y}) \in \mathbb{R} \times \mathbb{R}^n$, the following condition holds as $m \to \infty$:

$$\left| P\left(\frac{1}{\sqrt{m}}S_0([m-r]) \le x, \ \frac{1}{\sqrt{m}}S_m([mt_1]) \le y_1, \ \dots, \ \frac{1}{\sqrt{m}}S_m([mt_n]) \le y_n\right)\right.$$

$$\left. - P\left(\frac{1}{\sqrt{m}}S_0([m-r]) \le x\right)P\left(\frac{1}{\sqrt{m}}S_m([mt_i]) \le y_i \ \text{for all} \ i = 1, \dots, n\right)\right| \to 0,$$

$$(6)$$

where $r = r(m) = m^{1/(2+\delta)}$ and $[\cdot]$ denotes the integer part. This asymptotic independence relation ensures that, after removing a block of length $r$, the remaining partial sum is asymptotically independent of the finite-dimensional distributions of the partial sum process from the monitoring sequence.

Furthermore, we assume that there is a positive-definite $p \times p$ matrix $\mathbf{C}$ and a constant $\tau > 0$ such that

$$\left|\frac{1}{\ell}\sum_{i=1}^{\ell}\mathbf{x}_i\mathbf{x}_i^T - \mathbf{C}\right| = O\left(\ell^{-\tau}\right) \quad \text{a.s.} \quad (\ell \to \infty), \tag{7}$$

where $|\cdot|$ denotes the maximum norm of a matrix or vector, respectively. We also assume that

$$\left|\sum_{j=1}^{m}\mathbf{x}_j\varepsilon_j\right| = O_P\left(m^{1/2}\right) \quad (m \to \infty). \tag{8}$$

With the parameters $\sigma$ and $\tau$ introduced above we define the boundary function

$$g_m(k) = \sigma c g_m^*(k) = \sigma c m^{1/2}\left(1 + \frac{k}{m}\right)\left(\frac{k}{m+k}\right)^\gamma, \quad 0 \le \gamma < \min\{\tau, 1/2\}, \tag{9}$$

where $\gamma$ is a certain tuning constant.

We now state our main results.

**Theorem 2.1.** *Assume that conditions (1) - (9) hold. Then, under $H_0$, we have*

$$\lim_{m \to \infty} P\left(\frac{1}{\sigma}\sup_{1 \le k < \infty}\frac{|\hat{Q}_m(k)|}{g_m^*(k)} > c\right) = P\left(\sup_{0 < t \le 1}\frac{|W(t)|}{t^\gamma} > c\right),$$

*where $\{W(t), 0 \le t < \infty\}$ is a standard Wiener process.*

*Remark 0.1.*   Selected quantiles of the limiting distribution in Theorem 2.1 are given in [8]. Moreover, according to [9], in the case of $\gamma = 1/2$, which is excluded above, an asymptotic extreme value distribution can be derived via proving a Darling-Erdős type limit theorem.

An application of Theorem 2.1 in practice requires the estimation of the unknown parameter $\sigma$. This can be achieved by an application of the weak IP in (4). Let $\{\ell_m, 1 \le m < \infty\}$ be a non-decreasing sequence of positive integers with $1 \le \ell_m \le m$ such that $\lim_{m \to \infty} \ell_m/m = 0$ and

$$\liminf_{m \to \infty} \frac{\ell_m}{m^\theta} > 0 \text{ for some } \theta \text{ with } \max\left\{1 - 2\tau, 1 - \tau - \frac{\delta}{2(2+\delta)}\right\} < \theta < 1. \quad (10)$$

We propose the estimator

$$\widehat{\sigma}_m^2 = \frac{1}{k} \sum_{j=1}^{k} \left\{ \frac{1}{\sqrt{\ell}} \sum_{i=(j-1)\ell+1}^{j\ell} \widehat{\varepsilon}_i \right\}^2, \quad (11)$$

where $\ell = \ell_m$ and $k = k_m = [m/\ell]$.

*Remark 0.2.* The statistic above is not the same but motivated by the class of consistent estimators introduced in [16] for the case of a $\rho$-mixing sequence. Our approach here is to prove the consistency of $\widehat{\sigma}_m^2$ from (11) in the case of our general setting solely via the approximating Wiener process.

**Theorem 2.2.** *Assume that conditions* (1) - (11) *hold. Then, under $H_0$, we have*

$$\lim_{m \to \infty} P\left( \frac{1}{\widehat{\sigma}_m} \sup_{1 \le k < \infty} \frac{\left|\widehat{Q}_m(k)\right|}{g_m^*(k)} > c \right) = P\left( \sup_{0 < t \le 1} \frac{|W(t)|}{t^\gamma} > c \right),$$

*where $\{W(t), 0 \le t < \infty\}$ is a standard Wiener process.*

**Theorem 2.3.** *If $c_1$ denotes the first column of $C$, let $c_1^T (\beta_0 - \beta_*) \ne 0$. Assume that conditions* (1) - (11) *hold. Then, under $H_A$, we have*

$$\lim_{m \to \infty} P\left( \frac{1}{\widehat{\sigma}_m} \sup_{1 \le k < \infty} \frac{\left|\widehat{Q}_m(k)\right|}{g_m^*(k)} > M \right) = 1 \quad \text{for all} \quad M > 0.$$

# 3 Examples

## 3.1 Linear models with NED regressors

In this subsection it is shown that our assumptions hold for near-epoch dependent (NED) regressors. The NED concept covers widely used nonlinear time series like, e.g., GARCH models (cf. Davidson [7]). In the particular case of an NED sequence on an independent process, Ling [14] established a strong law of large numbers (SLLN) and a strong IP. For convenience we state the definition adopted from Ling [14].

Let $\{\varepsilon_i, -\infty < i < \infty\}$ be a sequence of independent random variables and $x_i$ be $\mathscr{F}_{-\infty}^i$-measurable, where $\mathscr{F}_k^\ell$ denotes the $\sigma$-field generated by $\{\varepsilon_j, k \le j \le \ell\}$.

**Definition 0.1.** The process $\{x_i, -\infty < i < \infty\}$ will be called $L_p(v)$-near-epoch-dependent $(L_p(v)$-NED) on $\{\varepsilon_i, -\infty < i < \infty\}$ if

$$\sup_{-\infty < i < \infty} \|x_i\|_p < \infty \quad \text{and} \quad \sup_{-\infty < i < \infty} \left\| x_i - E\left(x_i | \mathscr{F}^i_{i-k}\right) \right\|_p = O\left(k^{-\nu}\right) \quad (k \to \infty),$$

where $\|x\|_p = \left(E|x|^p\right)^{1/p}$, $p \geq 1$ and $\nu > 0$.

If $\{\varepsilon_i, -\infty < i < \infty\}$ is i.i.d. and centered with $\|\varepsilon_1\|_4 < \infty$ and $\{x_i, -\infty < i < \infty\}$ is an $L_4(\nu)$-NED sequence with $\nu > 1/2$ and constant variance $\sigma_x^2 > 0$, then

$$y_i = \beta_1 + \beta_2 x_{i-1} + \varepsilon_i, \quad i = 1, 2, \ldots,$$

allows for an application of the monitoring procedure, provided (4) - (8) hold.
We can establish (4) and (5) by the well-known Hungarian construction of Komlós, Major and Tusnády, [cf., e.g., 6, Theorem 2.6.3]. Condition (6) is clearly fulfilled. By Hölder's inequality, we get

$$\left\| x_{i-1} \varepsilon_i - E\left(x_{i-1} \varepsilon_i | \mathscr{F}^i_{i-k+1}\right) \right\|_2 \leq \|\varepsilon_i\|_4 \left\| x_{i-1} - E\left(x_{i-1} | \mathscr{F}^i_{i-k+1}\right) \right\|_4,$$

which implies that $\{x_{i-1} \varepsilon_i, -\infty < i < \infty\}$ is $L_2(\nu)$-NED. Moreover, using the conditional Jensen inequality and iterated conditional expectation, we have

$$\sup_{-\infty < i < \infty} \left\| E\left(x_{i-1} \varepsilon_i | \mathscr{F}^{i-k}_{-\infty}\right) \right\|_2 = O\left(k^{-\nu}\right) \quad (k \to \infty).$$

Thus, an application of McLeish's maximal inequality on mixingales [cf. 15] yields assumption (8). Since, due to the projection property of conditional expectations,

$$\left\| x_i^2 - E\left(x_i^2 | \mathscr{F}^i_{i-k+1}\right) \right\|_2 \leq \left\| x_i^2 - \left(E\left(x_i | \mathscr{F}^i_{i-k+1}\right)\right)^2 \right\|_2$$

holds, it can be shown that $\{x_i^2 - \sigma_x^2, -\infty < i < \infty\}$ is a centered $L_2(\nu)$-NED sequence. Then the desired rate in assumption (7) follows from the SLLN for NED processes obtained by [14].

*Remark 0.3.* Recently, Aue et al. [3] derived the limit distribution for the delay time of the CUSUM monitoring procedure. They extended the linear model to multiple regression models with time series regressors, i.e. AR-GARCH models, which also satisfy the NED property.

## 3.2 Linear models with asymptotically M-dependent errors

If the error sequence is $M$-dependent, i.e. independent at lags exceeding $M$, then (6) clearly holds. For our needs it suffices, if the error sequence can be approximated by an $M$-dependent sequence $\{\widetilde{\varepsilon}_i, 1 \leq i < \infty\}$, such that

$$\sup_{1 \leq k < \infty} k^{-1/(2+\delta)} \left| \sum_{i=m+1}^{m+k} \varepsilon_i - \sum_{i=m+1}^{m+k} \widetilde{\varepsilon}_i \right| = O_P(1) \quad (m \to \infty).$$

Then we can replace the partial sums from the monitoring period by the partial sums from these $M$-dependent random variables. We illustrate this for the augmented GARCH(1,1) process. Let $\{\varepsilon_i,\ 1 \leq i < \infty\}$, satisfy the equation

$$\varepsilon_i = \sigma_i \zeta_i,$$

together with the recursion

$$\Lambda\left(\sigma_i^2\right) = h(\zeta_{i-1})\Lambda\left(\sigma_{i-1}^2\right) + g(\zeta_{i-1}),$$

where $\{\zeta_i,\ i \in \mathbb{Z}\}$ is a centered i.i.d. sequence and $\Lambda(\cdot)$, $h(\cdot)$ and $g(\cdot)$ are suitable real-valued functions. If $\Lambda^{-1}(x)$ exists, Aue et al. [1] proved that

$$\Lambda\left(\sigma_i^2\right) = \sum_{j=1}^{\infty} g\left(\zeta_{i-j}\right)\prod_{k=1}^{j-1} h(\zeta_{i-k})$$

is the only stationary solution of the recursion. Under an additional smoothness condition on $\Lambda(\cdot)$, they also obtained a strong IP for the partial sums of the random variables $\varepsilon_i^2 - E\varepsilon_i^2$. This gives rise to a CUSUM monitoring with squared augmented GARCH(1,1) errors. Again, the crucial point is to ensure condition (6). This can be achieved using a construction along the lines of the proofs in Aue et al. [1]. This construction yields random variables

$$\widetilde{\varepsilon}_i \in \sigma\left(\zeta_j : i - i^\rho \leq j \leq i\right),$$

for some $0 < \rho < 1/(2+\delta)$. Although the lag depends on the index $i$, the lag at index $m$ suffices for condition (6), because the desired approximation above follows from Lemma 5.4 in Aue et al. [1]. For further details we refer to the proofs in Aue et al. [2]. Therein, additionally to the CUSUM monitoring, they introduced another monitoring procedure based on squared prediction errors.

## 3.3 Monitoring strongly mixing AR models

In order to measure the possible dependency of the error sequence $\{\varepsilon_i,\ 1 \leq i < \infty\}$, we now follow the strong mixing concept. The dependency of two sub-$\sigma$-fields $\mathcal{G}$ and $\mathcal{H}$ is measured by

$$\alpha\left(\mathcal{G}, \mathcal{H}\right) = \sup\left\{|P\left(A \cap B\right) - P\left(A\right)P\left(B\right)|\ A \in \mathcal{G},\ B \in \mathcal{H}\right\}.$$

As above, for consecutive integers $k$ and $\ell$ the notation $\mathcal{F}_k^\ell$ denotes the $\sigma$-field generated by $\{\varepsilon_j,\ k \leq j \leq \ell\}$. The (so-called) mixing coefficient $\alpha(n)$ is defined as

$$\alpha(n) = \sup_{p \in \mathbb{N}} \alpha\left(\mathcal{F}_1^p, \mathcal{F}_{p+n}^\infty\right) \quad \text{for } n = 1, 2, \ldots.$$

The sequence is called strongly mixing if $\lim_{n\to\infty}\alpha(n) = 0$, which clearly implies condition (6).

Now, in contrast to the monitoring procedure described in Horváth et al. [8], in this subsection we do not assume any longer that the regressors are independent of the errors. This, for example, allows for lag dependent variables as regressors. Consider, e.g., the direct application of the monitoring procedure to the AR(1) model $y_i = \beta y_{i-1} + u_i$. Since the mean regressor is zero the first assumption of Theorem 2.3 is not satisfied. But for detecting a change in the parameter of the AR(1) model, it suffices to monitor the linear model

$$y_i y_{i-1} = \begin{cases} \beta y_{i-1}^2 + u_i y_{i-1}, & m+1 \leq i < m+k^* \\ \beta_* y_{i-1}^2 + u_i y_{i-1}, & i = m+k^*, m+k^*+1, \ldots, \end{cases} \tag{12}$$

where $0 \leq \beta \neq \beta_* < 1$. In the sequel we assume that

$$\{u_i, -\infty < i < \infty\} \text{ is a centered i.i.d. sequence} \tag{13}$$

such that

$$\{y_i, 0 \leq i < \infty\} \text{ is strictly stationary, strongly mixing and} \atop \text{that condition (4) holds with } \varepsilon_i \text{ replaced by } u_i y_{i-1}. \tag{14}$$

Since $\{y_i, 0 \leq i < \infty\}$ is assumed to be a strictly stationary process, we mention that condition (14) also yields the uniform weak IP on the monitoring sequence.

If $\gamma(j) = Ey_1 y_{1+j}$ denotes the autocovariance function, we additionally assume that there is a positive constant $\tau > 0$ such that

$$\left| \frac{1}{m} \sum_{i=1}^{m} y_i^2 - \gamma(0) \right| = O\left(m^{-\tau}\right) \quad \text{a.s.} \quad (m \to \infty). \tag{15}$$

We note that $\{y_i^2 - \gamma(0), 0 \leq i < \infty\}$ is also strongly mixing. Thus, it is possible to replace (15) by a moment condition such that the squares obey an invariance principle [cf., e.g., 13, Theorem 4]. Then, via the approximating Wiener process, a Marcinkiewicz-Zygmund type law of large numbers yields the desired assertion.

We set $\hat{\beta}_m = \{\sum_{i=0}^{m-1} y_i^2\}^{-1} \sum_{i=0}^{m-1} y_i y_{i+1}$ and $\hat{R}_m(k) = \sum_{i=m+1}^{m+k} (y_i y_{i-1} - \hat{\beta}_m y_{i-1}^2)$.

**Theorem 3.1.** *Assume that conditions (9) and (12) - (15) hold. Then, under $H_0$, we have*

$$\lim_{m\to\infty} \mathbf{P}\left( \frac{1}{\Gamma} \sup_{1 \leq k < \infty} \frac{|\hat{R}_m(k)|}{g_m^*(k)} > c \right) = \mathbf{P}\left( \sup_{0 < t \leq 1} \frac{|W(t)|}{t^\gamma} > c \right),$$

*where $\Gamma^2 = E(u_1 y_0)^2$ and $\{W(t), 0 \leq t < \infty\}$ is a standard Wiener process.*

The statement of Theorem 3.1 remains true if we plug in a consistent estimator $\hat{\Gamma}_m$ based on the non-contaminated training period. According to condition (14), $\Gamma$ appears in the uniform weak IP, i.e. $\sum_{i=1}^{m} u_i y_{i-1} - \Gamma W(m) = O_P\left(m^{1/(2+\delta)}\right)$. Computing the covariance of the AR(1) process yields $\sigma_u^2 = Eu_1^2 = \gamma(0) - \beta\gamma(1)$. Therefore the natural estimator $\widehat{\Gamma}_m^2 = \widehat{\gamma}_m^2(0) - \widehat{\beta}_m\widehat{\gamma}_m(1)\widehat{\gamma}_m(0)$ is a combination of covariance estimators and the least squares estimator $\widehat{\beta}_m$ and hence is consistent.

Under the alternative, i.e. $k^* < \infty$ in (12), we additionally assume that the "post-change" AR process

$$\{y_{m+k^*+i}, \ 0 \le i < \infty\} \text{ is strongly mixing, satisfies condition (15),}$$
$$\text{and that condition (4) holds with } \varepsilon_i \text{ replaced by } u_{m+k^*+i}y_{m+k^*+i-1}. \tag{16}$$

Then we can establish asymptotic power one of the monitoring procedure.

**Theorem 3.2.** *Assume that conditions (9) and (12) - (16) hold. Then, under $H_A$, we have*

$$\lim_{m \to \infty} \mathbf{P}\left( \frac{1}{\widehat{\Gamma}_m} \sup_{1 \le k < \infty} \frac{\left|\widehat{R}_m(k)\right|}{g_m^*(k)} > M \right) = 1 \quad \text{for all} \quad M > 0.$$

*Remark 0.4.* We point out that the monitoring of (12) overcomes the restriction in Theorem 2.3 concerning detectable parameter shifts. For another approach we refer to [10] and [12] who introduced CUSUM type test statistics based on weighted residuals which are able to detect any change in the slope parameter of a linear model with independent errors. Moreover, they extended these results to AR(p) time series [cf. 11].

# 4 Proofs

We first prove a lemma, which provides a general result on the asymptotic independence of stochastic processes. In the following, let $T$ be a positive integer and denote $D = D\left[T^{-1}, T\right]$ to be the space of real functions on $\left[T^{-1}, T\right]$ which are right-continuous and have left-hand limits. The space $D$ is endowed with the Skorohod metric $s$ [cf. 4]. Moreover, we set $D^2 = D \times D$ to be the product space equipped with the metric $s^{(2)}$, i.e., $s^{(2)}\left((f_1, f_2), (g_1, g_2)\right) = s(f_1, g_1) \vee s(f_2, g_2)$. The law of a random element $f$ of $D$ is denoted by $\mathscr{L}(f)$ and $\xrightarrow{D}$ means weak convergence in the metric space $D$. We write $\Rightarrow$ for the weak convergence of finite-dimensional random vectors. Now we fix a positive integer $n$ and define, for each $T^{-1} = t_1 < \cdots < t_n = T$, the function $\pi_n$ from $D$ to $\mathbb{R}^n$ by $\pi_n(f) = (f(t_1), \dots, f(t_n))$. Let $[t]$ denote the integer part of $t$ and, for the sake of convenience, set $(t)_m = ([mt] + 1)/m$. Finally, we define two more mappings, i.e. $id : [T^{-1}, T] \to [T^{-1}, T], \ t \mapsto t$, and $h : [T^{-1}, T] \to \mathbb{R}, \ t \mapsto (1+t)(t/(1+t))^\gamma$.

**Lemma 4.1.** *Let $\{(f_m, g_m), \ 1 \le m < \infty\}$ and $(f, g)$ be $D^2$-valued random elements which are measurable with respect to the Borel $\sigma$-field $\mathscr{B}(D^2)$. Suppose that*

$$\mathscr{L}(f_m) \xrightarrow{D} \mathscr{L}(f) \quad \text{and} \quad \mathscr{L}(g_m) \xrightarrow{D} \mathscr{L}(g) \quad (m \to \infty), \tag{17}$$

*and, for each n,*

$$\mathscr{L}\left(\pi_n\left(f_m\right),\pi_n\left(g_m\right)\right) \overset{\mathbb{R}^{2n}}{\Rightarrow} \mathscr{L}\left(\pi_n\left(f\right)\right)\otimes\mathscr{L}\left(\pi_n\left(g\right)\right) \quad (m\to\infty). \tag{18}$$

*Then we have*

$$\mathscr{L}\left(f_m,g_m\right) \overset{D}{\to} \mathscr{L}\left(f\right)\otimes\mathscr{L}\left(g\right) \quad (m\to\infty). \tag{19}$$

*Proof.* We choose $A_n = j_n \circ \pi_n$, where $j_n : \mathbb{R}^n \to D$ maps the vector $(\xi_1,\dots,\xi_n)$ onto the function whose value in $t_n$ is $\xi_n$ and which is constant $\xi_i$ on the interval $[t_i,t_{i+1})$, for each $1 \le i \le n-1$. A combination of the continuous mapping theorem and (18) yields

$$\mathscr{L}\left(A_n\left(f_m\right),A_n\left(g_m\right)\right) \overset{D^2}{\to} \mathscr{L}\left(A_n\left(f\right)\right)\otimes\mathscr{L}\left(A_n\left(g\right)\right) \quad \text{for each } n. \tag{20}$$

By the definition of the Skorohod metric, we have

$$s\left(f,A_n\left(f\right)\right) \le \sup_{1\le i\le n}\ \sup_{t\in[t_i,t_i+1)}\ \left|f\left(t\right)-f\left(t_i\right)\right|.$$

Hence, from the right-continuity of $(f,g)$ together with condition (17),

$$\lim_{n\to\infty}\limsup_{m\to\infty} P\left(s^{(2)}\left(\left(f_m,g_m\right),\left(A_n\left(f_m\right),A_n\left(g_m\right)\right)\right) > \delta\right) = 0 \tag{21}$$

and

$$\lim_{n\to\infty} P\left(s^{(2)}\left(\left(f,g\right),\left(A_n\left(f\right),A_n\left(g\right)\right)\right) > \delta\right) = 0 \tag{22}$$

for each $\delta > 0$. In view of (20), (21) and (22), Proposition 1 in Wichura [20] gives (19) and thus completes the proof.

**Lemma 4.2.** *Under the conditions of Theorem 2.1, we have, as $m \to \infty$,*

$$\mathscr{L}\left(\frac{(\cdot)_m}{g_m^*\left(m(\cdot)_m\right)}W_{0,m}\left(m\right)\right) \overset{D}{\to} \mathscr{L}\left(\frac{id(\cdot)}{h(\cdot)}W\left(1\right)\right) \tag{23}$$

*and*

$$\mathscr{L}\left(\frac{1}{g_m^*\left(m(\cdot)_m\right)}W_{1,m}\left(m(\cdot)_m\right)\right) \overset{D}{\to} \mathscr{L}\left(\frac{1}{h(\cdot)}W\left(\cdot\right)\right). \tag{24}$$

*Proof.* By the scaling property of the Wiener process, we observe

$$\left\{\frac{1}{g_m^*\left(m(t)_m\right)}W_{1,m}\left(m(t)_m\right),\ T^{-1}\le t\le T\right\}$$

$$\overset{d}{=} \left\{\frac{1}{h\left((t)_m\right)}W\left((t)_m\right),\ T^{-1}\le t\le T\right\}. \tag{25}$$

Moreover, the almost sure continuity of $W$ gives

$$\lim_{m \to \infty} P \left( \sup_{T^{-1} \le t \le T} \left| \frac{1}{h((t)_m)} W((t)_m) - \frac{1}{h(t)} W(t) \right| > \delta \right) = 0 \qquad (26)$$

for each $\delta > 0$. A combination of (25) and (26) yields (24). Using similar arguments we also arrive at (23).

**Lemma 4.3.** *Under the conditions of Theorem 2.1, we have, as* $m \to \infty$,

$$\mathcal{L} \left( \pi_n \left( \cdot \frac{1}{g_m^*(m(\cdot)_m)} W_{1,m}(m(\cdot)_m) \right), \pi_n \left( \frac{(\cdot)_m}{g_m^*(m(\cdot)_m)} W_{0,m}(m) \right) \right)$$

$$\overset{\mathbb{R}^{2n}}{\Rightarrow} \mathcal{L} \left( \pi_n \left( \frac{1}{h(\cdot)} W(\cdot) \right) \right) \otimes \mathcal{L} \left( \pi_n \left( \frac{id(\cdot)}{h(\cdot)} W(1) \right) \right). \qquad (27)$$

*Proof.* Throughout this proof we set $r = r(m) = m^{1/(2+\delta)}$. From (4) together with a standard estimate on the increments of the Wiener process [cf. 6, Theorem 1.2.1] we get

$$S_0(m) - S_0([m-r]) = O_P \left( m^{1/(4+2\delta)} \sqrt{\log m} \right) + O_P \left( m^{1/(2+\delta)} \right)$$

$$= O_P \left( m^{1/(2+\delta)} \right) \quad (m \to \infty). \qquad (28)$$

Thus a combination of (4) and (28) results in

$$\frac{(t_k)_m}{g_m^*(m(t_k)_m)} \left( \sigma W_{0,m}(m) - S_0([m-r]) \right) = o_P(1) \quad (m \to \infty), \qquad (29)$$

for each $1 \le k \le n$. We further note that (5) yields

$$\frac{1}{g_m^*(m(t_k)_m)} \left( \sigma W_{1,m}(m(t_k)_m) - S_m(m(t_k)_m) \right) = o_P(1) \quad (m \to \infty). \qquad (30)$$

According to (29) and (30) it is enough to establish the asymptotic independence for the approximating partial sum processes. Let $\boldsymbol{x}, \boldsymbol{y} \in \mathbb{R}^n$ and observe, as $m \to \infty$,

$$\left\{ \pi_n \left( \frac{(\cdot)_m}{g_m^*(m(\cdot)_m)} S_0([m-r]) \right) \le \boldsymbol{x} \right\}$$

$$= \left\{ m^{-1/2} S_0([m-r]) \le \min_{1 \le k \le n} \left( \frac{h(t_k)}{t_k} x_k \right) + o(1) \right\} \qquad (31)$$

and

$$\left\{ \pi_n \left( \frac{1}{g_m^*(m(\cdot)_m)} S_m(m(\cdot)_m) \right) \le \boldsymbol{y} \right\}$$

$$= \left\{ m^{-1/2} S_m([mt_1]) + m^{-1/2} \varepsilon_{m+[mt_1]+1} \le h(t_1) y_1 + o(1), \qquad (32) \right.$$

$$\left. \dots, m^{-1/2} S_m([mt_n]) + m^{-1/2} \varepsilon_{m+[mt_n]+1} \le h(t_n) y_n + o(1) \right\}.$$

By assumption (5), we clearly have $m^{-1/2}\varepsilon_{m+[mt_k]+1} = o_P(1)$, as $m \to \infty$, uniformly in $1 \le k \le n$. For the remaining partial sums on the right-hand side of (31) and (32), via (4) and (5), the central limit theorem yields absolutely continuous marginal limit distributions. Therefore,

$$\left| P\left\{ \pi_n \left( \frac{1}{g_m^*(m(\cdot)_m)} W_{1,m}(m(\cdot)_m) \right) \le x, \; \pi_n \left( \frac{(\cdot)_m}{g_m^*(m(\cdot)_m)} W_{0,m}(m) \right) \le y \right\} \right.$$

$$\left. - P\left\{ m^{-1/2} S_0([m-r]) \le \min_{1 \le k \le n} \left( \frac{h(t_k)}{t_k} x_k \right), \; m^{-1/2} S_m([mt_1]) \le h(t_1) y_1, \right. \right.$$

$$\left. \left. \ldots, m^{-1/2} S_m([mt_n]) \le h(t_n) y_n \right\} \right| = o(1) \quad (m \to \infty).$$

The asymptotic independence now follows directly from (6), and an application of (23) and (24) completes the proof.

**Lemma 4.4.**  *Under the conditions of Theorem 2.1, we have, as $m \to \infty$,*

$$\sup_{1+[mT^{-1}] \le k \le mT+1} \frac{1}{g_m^*(k)} \left| W_{1,m}(k) - \frac{k}{m} W_{0,m}(m) \right|$$

$$\Rightarrow \sup_{(1+T)^{-1} \le s \le T/(1+T)} \frac{|W(s)|}{s^\gamma}. \tag{33}$$

*Proof.*  Observe that

$$\sup_{1+[mT^{-1}] \le k \le mT+1} \frac{1}{g_m^*(k)} \left| W_{1,m}(k) - \frac{k}{m} W_{0,m}(m) \right|$$

$$= \sup_{T^{-1} \le t \le T} \frac{1}{g_m^*(m(t)_m)} \left| W_{1,m}(m(t)_m) - (t)_m W_{0,m}(m) \right|. \tag{34}$$

Now, Lemmas 4.1 - 4.3 together with the continuous mapping theorem yield

$$\sup_{T^{-1} \le t \le T} \frac{1}{g_m^*(m(t)_m)} \left| W_{1,m}(m(t)_m) - (t)_m W_{0,m}(m) \right|$$

$$\Rightarrow \sup_{T^{-1} \le t \le T} \frac{1}{h(t)} \left| \tilde{W}_1(t) - t \tilde{W}_0(1) \right| \quad (m \to \infty), \tag{35}$$

where $\left\{ \tilde{W}_1(t), \; 0 \le t < \infty \right\}$ and $\left\{ \tilde{W}_0(t), \; 0 \le t < \infty \right\}$ are standard Wiener processes such that $\sigma\left( \tilde{W}_1(t), \; 0 \le t < \infty \right)$ and $\sigma\left( \tilde{W}_0(1) \right)$ are independent $\sigma$-fields. According to Horváth et al. [8] this implies

$$\left\{ \tilde{W}_1(t) - t\tilde{W}_0(1), \; 0 \le t < \infty \right\} \overset{d}{=} \left\{ (1+t) W(t/(1+t)), \; 0 \le t < \infty \right\}, \tag{36}$$

where $\{W(t),\ 0 \leq t < \infty\}$ is a standard Wiener process again. Therefore a combination of (34) - (36) results in (33).

*Proof (of Theorem 2.1).*  In view of (2) and (7) - (9), Lemma 5.2 in Horváth et al. [8] gives

$$\sup_{1 \leq k < \infty} \frac{1}{g_m^*(k)} \left| \sum_{i=m+1}^{m+k} \widehat{\varepsilon}_i - \left( \sum_{i=m+1}^{m+k} \varepsilon_i - \frac{k}{m} \sum_{i=1}^{m} \varepsilon_i \right) \right| = o_P(1) \quad (m \to \infty). \tag{37}$$

Moreover, under (4), (5) and (9), following the proof of Lemma 5.3 in Horváth et al. [8] we get

$$\begin{aligned} \sup_{1 \leq k < \infty} \frac{1}{g_m^*(k)} &\left| \left( \sum_{i=m+1}^{m+k} \varepsilon_i - \frac{k}{m} \sum_{i=1}^{m} \varepsilon_i \right) \right. \\ &\left. - \left( W_{1,m}(k) - \frac{k}{m} W_0(m) \right) \right| = o_P(1) \quad (m \to \infty). \end{aligned} \tag{38}$$

Due to the approximations (37) and (38), the law of the iterated logarithm for the Wiener process at zero and at infinity yields

$$\begin{aligned} \lim_{T \to \infty} \limsup_{m \to \infty} P &\left( \left| \sup_{1 \leq k < \infty} \frac{1}{g_m^*(k)} \left| \widehat{Q}_m(k) \right| \right. \right. \\ &\left. \left. - \max_{1+[mT^{-1}] \leq t \leq mT+1} \frac{1}{g_m^*(k)} \left| \widehat{Q}_m(k) \right| \right| > \delta \right) = 0 \end{aligned} \tag{39}$$

for each $\delta > 0$. For details we refer to Horváth et al. [8] and Schmitz [18]. Similarly,

$$\lim_{T \to \infty} P \left( \left| \sup_{0 < t \leq 1} \frac{|W(t)|}{t^{\gamma}} - \max_{(1+T)^{-1} \leq t \leq T/(1+T)} \frac{|W(t)|}{t^{\gamma}} \right| > \delta \right) = 0. \tag{40}$$

Now Theorem 2.1 follows from (37) - (40) together with Lemma 4.4.

*Proof (of Theorem 2.2).*  It is enough to show that $\widehat{\sigma}_m^2 - \sigma^2 = o_P(1)$ as $m \to \infty$. First note that (7) and (8) yield

$$\left| (\widehat{\boldsymbol{\beta}}_m - \boldsymbol{\beta}_0) - \frac{1}{m} \mathbf{C}^{-1} \sum_{i=1}^{m} \mathbf{x}_i \varepsilon_i \right| = O_P \left( m^{-\frac{1}{2}-\tau} \right) \quad (m \to \infty).$$

If $\mathbf{c}_1$ denotes the first column of $\mathbf{C}$, by (2) we have that $\mathbf{c}_1^T \mathbf{C}^{-1} = (1,0,\ldots,0)$ and therefore, as $m \to \infty$,

$$\sum_{i=(j-1)\ell+1}^{j\ell} \mathbf{x}_i^T \left(\widehat{\boldsymbol{\beta}}_m - \boldsymbol{\beta}_0\right) = \ell \mathbf{c}_1^T \left(\widehat{\boldsymbol{\beta}}_m - \boldsymbol{\beta}_0\right) + O_P\left(m^{\frac{1}{2}-2\tau}\right)$$

$$= \frac{\ell}{m} \mathbf{c}_1^T \mathbf{C}^{-1} \sum_{i=1}^{m} \mathbf{x}_i \varepsilon_i + O_P\left(m^{\frac{1}{2}-\tau}\right) + O_P\left(m^{\frac{1}{2}-2\tau}\right)$$

$$= \frac{\ell}{m} \sum_{i=1}^{m} \varepsilon_i + O_P\left(m^{\frac{1}{2}-\tau}\right).$$

Hence

$$\frac{1}{\ell}\left(\sum_{i=(j-1)\ell+1}^{j\ell} \widehat{\varepsilon}_i\right)^2 - \frac{1}{\ell}\left(\sum_{i=(j-1)\ell+1}^{j\ell} \varepsilon_i - \frac{\ell}{m}\sum_{i=1}^{m} \varepsilon_i\right)^2$$

$$= \frac{1}{\ell}O_P\left(m^{\frac{1}{2}-\tau}\right)\left(2\sum_{i=(j-1)\ell+1}^{j\ell} \varepsilon_i - 2\frac{\ell}{m}\sum_{i=1}^{m} \varepsilon_i + O_P\left(m^{\frac{1}{2}-\tau}\right)\right). \tag{41}$$

Next, by (4) and a standard estimate on the increments of a Wiener process [cf. 6, Theorem 1.2.1], we get

$$\sum_{i=(j-1)\ell+1}^{j\ell} \varepsilon_i = \sum_{i=1}^{j\ell} \varepsilon_i - \sum_{i=1}^{(j-1)\ell} \varepsilon_i$$

$$= O_P\left(m^{1/(2+\delta)}\right) + O_P\left(\sqrt{\ell \log(m/\ell)}\right) \quad (m \to \infty). \tag{42}$$

In view of (41), (42) and (10), the left-hand side of (41) is stochastically bounded of order

$$O_P\left(\sqrt{\log(m/\ell)}\,\frac{m^{\frac{1}{2}-\tau}}{\sqrt{\ell}} + \frac{m}{\ell}\left(m^{-\tau-\frac{\delta}{2(2+\delta)}} + m^{-2\tau}\right)\right) = o_P(1) \quad (m \to \infty),$$

and therefore,

$$\widehat{\sigma}_m^2 = \frac{1}{k}\sum_{j=1}^{k}\frac{1}{\ell}\left(\sum_{i=(j-1)\ell+1}^{j\ell} \varepsilon_i - \frac{\ell}{m}\sum_{i=1}^{m} \varepsilon_i\right)^2 + o_P(1) \quad (m \to \infty). \tag{43}$$

Finally, an application of Theorem 2.1 in [19] completes the proof.

*Proof (of Theorem 2.3).* Set $\widetilde{k} = k^* + m$. Then

$$\sum_{i=m+1}^{m+\widetilde{k}} \widehat{\varepsilon}_i = \sum_{i=m+1}^{m+\widetilde{k}} \varepsilon_i + \left(\sum_{i=m+1}^{m+\widetilde{k}} \mathbf{x}_i\right)^T (\boldsymbol{\beta}_0 - \widehat{\boldsymbol{\beta}}_m)$$

$$+ \left(\sum_{i=m+k^*+1}^{m+\widetilde{k}} \mathbf{x}_i\right)^T (\boldsymbol{\beta}_* - \boldsymbol{\beta}_0)$$

$$= A_m + B_m,$$

where $B_m$ denotes the last term. Theorem 2.1 yields that $|A_m|/g_m^*(\widetilde{k}) = O_P(1)$, as $m \to \infty$. Since the mean regressor is not orthogonal to the shift $\boldsymbol{\beta}_* - \boldsymbol{\beta}_0$, one can show via (7) that $\liminf_{m\to\infty} |B_m|/\left\{mh(\widetilde{k}/m)\right\} > 0$ holds almost surely. For details we refer to the proof of Theorem 2.2 in Horváth et al. [8].

*Proof (of Theorem 3.1).* We have

$$\left|\widehat{R}_m(k)\right| = \left|\sum_{i=m+1}^{m+k} u_i y_{i-1} - \left\{\sum_{i=0}^{m-1} y_i^2\right\}^{-1}\left\{\sum_{i=0}^{m-1} u_{i+1}y_i\right\}\left\{\sum_{i=m+1}^{m+k} y_{i-1}^2\right\}\right|.$$

Hence, by (15), as $m \to \infty$,

$$\sup_{1 \le k < \infty} \frac{1}{g_m^*(k)} \left|\widehat{R}_m(k) - \sum_{i=m+1}^{m+k} u_i y_{i-1} + \frac{k}{m}\sum_{i=0}^{m-1} u_{i+1}y_i\right| = o_P(1),$$

and the proof can be completed in the same way as that of Theorem 2.1.

*Proof (of Theorem 3.2).* It follows similarly to the proof of Theorem 2.3.

**Acknowledgements** The authors wish to thank Dr. Mario Kühn for helpful comments. We also like to thank an anonymous referee for a number of suggestions that led to a significant improvement of the presentation.

# References

[1] Aue, A., Berkes, I., and Horváth, L. (2006a). Strong approximation for the sums of squares of augmented GARCH sequences. Bernoulli, 12, 583-608.

[2] Aue, A., Horváth, L., Hušková, M., and Kokoszka, P. (2006b). Change-point monitoring in linear models. Econometrics Journal, 9, 373-403.

[3] Aue, A., Horváth, L., and Reimherr, M. (2009). Delay times of sequential procedures for multiple time series regression models. Journal of Econometrics, 149, 174-190.

[4] Billingsley, P. (1968). Convergence of Probability Measures. Wiley, New York.

[5] Chu, C.-S. J., Stinchcombe, M., and White, H. (1996). Monitoring structural change. Econometrica, 64, 1045-1065.

[6] Csörgő, M. and Révész, P. (1981). Strong Approximations in Probability and Statistics. Academic Press, New York.

[7] Davidson, J. (2002). Establishing conditions for the functional central limit theorem in nonlinear time series processes. Journal of Econometrics, 106, 243-269.

[8] Horváth, L., Hušková, M., Kokoszka, P., and Steinebach, J. (2004). Monitoring changes in linear models. Journal of Statistical Planning and Inference, 126, 225-251.

[9] Horváth, L., Kokoszka, P., and Steinebach, J. (2007). On sequential detection of parameter changes in linear regression. Statistics and Probability Letters, 77, 885-895.

[10] Hušková, M. and Koubková, A. (2005). Monitoring jump changes in linear models. Journal of Statistical Research, 39, 51-70.

[11] Hušková, M. and Koubková, A. (2006). Sequential procedures for detection of changes in autoregressive sequences. Proceedings of "Prague Stochastics 2006" (eds. M. Hušková and M. Janžura), 437-447, MATFYZPRESS, Prague.

[12] Koubková, A. (2006). Sequential change-point analysis. PhD thesis, Charles University Prague.

[13] Kuelbs, J. and Philipp, W. (1980). Almost sure invariance principles for partial sums of mixing B-valued random variables. Annals of Probability, 8, 1003-1036.

[14] Ling, S. (2007). Testing for change points in time series models and limiting theorems for NED sequences. Annals of Statistics, 35, 1213-1237.

[15] McLeish, D.L. (1975). A maximal inequality and dependent strong laws. Annals of Probability, 5, 829-839.

[16] Peligrad, M. and Shao, Q.-M. (1995). Estimation of the variance of partial sums for $\rho$-mixing random variables. Journal of Multivariate Analysis, 52, 140-157.

[17] Perron, P. and Qu, Z. (2007). Estimating and testing multiple structural changes in multivariate regressions. Econometrica, 75, 459-502.

[18] Schmitz, A. (2007). Monitoring changes in dependent data, Diploma thesis. University of Cologne.

[19] Steinebach, J. (1995). Variance estimation based on invariance principles. Statistics, 27, 15-25.

[20] Wichura, M.J. (1971). A note on the weak convergence of stochastic processes. Annals of Mathematical Statistics, 42, 1769-1772.

# Testing for homogeneity of variance in the wavelet domain.

Olaf Kouamo, Eric Moulines, and Francois Roueff

**Abstract** The danger of confusing long-range dependence with non-stationarity has been pointed out by many authors. Finding an answer to this difficult question is of importance to model time-series showing trend-like behavior, such as river run-off in hydrology, historical temperatures in the study of climates changes, or packet counts in network traffic engineering.

The main goal of this paper is to develop a test procedure to detect the presence of non-stationarity for a class of processes whose $K$-th order difference is stationary. Contrary to most of the proposed methods, the test procedure has the same distribution for short-range and long-range dependence covariance stationary processes, which means that this test is able to detect the presence of non-stationarity for processes showing long-range dependence or which are unit root.

The proposed test is formulated in the wavelet domain, where a change in the generalized spectral density results in a change in the variance of wavelet coefficients at one or several scales. Such tests have been already proposed in [26], but these authors do not have taken into account the dependence of the wavelet coefficients within scales and between scales. Therefore, the asymptotic distribution of the test they have proposed was erroneous; as a consequence, the level of the test under the null hypothesis of stationarity was wrong.

In this contribution, we introduce two test procedures, both using an estimator of the variance of the scalogram at one or several scales. The asymptotic distribution of the test under the null is rigorously justified. The pointwise consistency of the test in the presence of a single jump in the general spectral density is also be presented. A limited Monte-Carlo experiment is performed to illustrate our findings.

O. Kouamo
ENSP, LIMSS, BP : 8390 Yaoundé, e-mail: olaf.kouamo@telecom-paristech.fr

E. Moulines, F. Roueff
Institut Télécom,
Télécom ParisTech, CNRS UMR 5181, Paris,
e-mail: eric.moulines,francois.roueff@telecom-paristech.fr

P. Doukhan et al. (eds.), *Dependence in Probability and Statistics*,
Lecture Notes in Statistics 200, DOI 10.1007/978-3-642-14104-1_10,
© Springer-Verlag Berlin Heidelberg 2010

# 1 Introduction

For time series of short duration, stationarity and short-range dependence have usually been regarded to be approximately valid. However, such an assumption becomes questionable in the large data sets currently investigated in geophysics, hydrology or financial econometrics. There has been a long lasting controversy to decide whether the deviations to "short memory stationarity" should be attributed to long-range dependence or are related to the presence of breakpoints in the mean, the variance, the covariance function or other types of more sophisticated structural changes. The links between non-stationarity and long-range dependence (LRD) have been pointed out by many authors in the hydrology literature long ago: [16] and [5] show that non-stationarity in the mean provides a possible explanations of the so-called Hurst phenomenon. [24] and later [25] suggested that more sophisticated changes may occur, and have proposed a method to detect such changes. The possible confusions between long-memory and some forms of nonstationarity have been discussed in the applied probability literature: [3] show that long-range dependence may be confused with the presence of a small monotonic trend. This phenomenon has also been discussed in the econometrics literature. [12] proposed a test of presence of structural change in a long memory environment. [11] showed that linear processes with breaks can mimic the autocovariance structure of a linear fractionally integrated long-memory process (a stationary process that encounters occasional regime switches will have some properties that are similar to those of a long-memory process). Similar behaviors are considered in [9] who provided simple and intuitive econometric models showing that long-memory and structural changes are easily confused. [18] asserted that what had been seen by many authors as long memory in the volatility of the absolute values or the square of the log-returns might, in fact, be explained by abrupt changes in the parameters of an underlying GARCH-type models. [2] proposed a testing procedure for distinguishing between a weakly dependent time series with change-points in the mean and a long-range dependent time series. [13] have proposed a test procedure for detecting long memory in presence of deterministic trends.

The procedure described in this paper deals with the problem of detecting changes which may occur in the spectral content of a process. We will consider a process $X$ which, before and after the change, is not necessary stationary but whose difference of at least a given order is stationary, so that polynomial trends up to that order can be discarded. Denote by $\Delta X$ the first order difference of $X$,

$$[\Delta X]_n \stackrel{\text{def}}{=} X_n - X_{n-1}, \quad n \in \mathbb{Z},$$

and define, for an integer $K \geq 1$, the $K$-th order difference recursively as follows: $\Delta^K = \Delta \circ \Delta^{K-1}$. A process $X$ is said to be $K$-th order difference stationary if $\Delta^K X$ is covariance stationary. Let $f$ be a non-negative $2\pi$-periodic symmetric function such that there exists an integer $K$ satisfying, $\int_{-\pi}^{\pi} |1 - e^{-i\lambda}|^{2K} f(\lambda) d\lambda < \infty$. We say that the process $X$ admits *generalized spectral density* $f$ if $\Delta^K X$ is weakly stationary and with spectral density function

$$f_K(\lambda) = |1 - e^{-i\lambda}|^{2K} f(\lambda).\tag{1}$$

This class of process include both short-range dependent and long-range dependent processes, but also unit-root and fractional unit-root processes. The main goal of this paper is to develop a testing procedure for distinguishing between a $K$-th order stationary process and a non-stationary process.

In this paper, we consider the so-called *a posteriori* or *retrospective* method (see [6, Chapter 3]). The proposed test is formulated in the wavelet domain, where a change in the generalized spectral density results in a change in the variance of the wavelet coefficients. Our test is based on a CUSUM statistic, which is perhaps the most extensively used statistic for detecting and estimating change-points in mean. In our procedure, the CUSUM is applied to the partial sums of the squared wavelet coefficients at a given scale or on a specific range of scales. This procedure extends the test introduced in [14] to detect changes in the variance of an independent sequence of random variables. To describe the idea, suppose that, under the null hypothesis, the time series is $K$-th order difference stationary and that, under the alternative, there is one breakpoint where the generalized spectral density of the process changes. We consider the scalogram in the range of scale $J_1, J_1 + 1, \ldots, J_2$. Under the null hypothesis, there is no change in the variance of the wavelet coefficients at any given scale $j \in \{J_1, \ldots, J_2\}$. Under the alternative, these variances takes different values before and after the change point. The amplitude of the change depends on the scale, and the change of the generalized spectral density. We consider the $(J_2 - J_1 + 1)$-dimensional W2-CUSUM statistic $\{T_{J_1,J_2}(t), t \in [0,1]\}$ defined by (41), which is a CUSUM-like statistics applied to the square of the wavelet coefficients. Using $T_{J_1,J_2}(t)$ we can construct an estimator $\widehat{\tau}_{J_1,J_2}$ of the change point (no matter if a change-point exists or not), by minimizing an appropriate norm of the W2-CUSUM statistics, $\widehat{\tau}_{J_1,J_2} = \mathrm{argmin}_{t \in [0,1]} \|T_{J_1,J_2}(t)\|_\star$. The statistic $T_{J_1,J_2}(\widehat{\tau}_{J_1,J_2})$ converges to a well-know distribution under the null hypothesis (see Theorems 3.1 and 3.2) but diverges to infinity under the alternative (Theorems 5.1 and 5.2). A similar idea has been proposed by [26] but these authors did not take into account the dependence of wavelet coefficient, resulting in an erroneous normalization and asymptotic distributions.

The paper is organized as follows. In Section 2, we introduce the wavelet setting and the relationship between the generalized spectral density and the variance of wavelet coefficients at a given scale. In Section 3, our main assumptions are formulated and the asymptotic distribution of the W2-CUSUM statistics is presented first in the single scale (sub-section 3.1) and then in the multiple scales (sub-section 3.2) cases. In Section 4, several possible test procedures are described to detect the presence of changes at a single scale or simultaneously at several scales. In Section 6, finite sample performance of the test procedure is studied based on Monte-Carlo experiments.

## 2 The wavelet transform of $K$-th order difference stationary processes

In this section, we introduce the wavelet setting, define the scalogram and explain how spectral change-points can be observed in the wavelet domain. The main advantage of using the wavelet domain is to alleviate problems arising when the time series exhibit is long range dependent. We will recall some basic results obtained in [19] to support our claims. We refer the reader to that paper for the proofs of the stated results.

**The wavelet setting.** The wavelet setting involves two functions $\phi$ and $\psi$ and their Fourier transforms

$$\widehat{\phi}(\xi) \stackrel{\text{def}}{=} \int_{-\infty}^{\infty} \phi(t) e^{-i\xi t} \, dt \quad \text{and} \quad \widehat{\psi}(\xi) \stackrel{\text{def}}{=} \int_{-\infty}^{\infty} \psi(t) e^{-i\xi t} \, dt,$$

and assume the following:

1. $\phi$ and $\psi$ are compactly-supported, integrable, and $\widehat{\phi}(0) = \int_{-\infty}^{\infty} \phi(t) \, dt = 1$ and $\int_{-\infty}^{\infty} \psi^2(t) \, dt = 1$.
2. There exists $\alpha > 1$ such that $\sup_{\xi \in \mathbb{R}} |\widehat{\psi}(\xi)| (1 + |\xi|)^{\alpha} < \infty$.
3. The function $\psi$ has $M$ vanishing moments, *i.e.* $\int_{-\infty}^{\infty} t^m \psi(t) \, dt = 0$ for all $m = 0, \ldots, M-1$
4. The function $\sum_{k \in \mathbb{Z}} k^m \phi(\cdot - k)$ is a polynomial of degree $m$ for all $m = 0, \ldots, M-1$.

The fact that both $\phi$ and $\psi$ have finite support (Condition 1) ensures that the corresponding filters (see (7)) have finite impulse responses (see (9)). While the support of the Fourier transform of $\psi$ is the whole real line, Condition 2 ensures that this Fourier transform decreases quickly to zero. Condition 3 is an important characteristic of wavelets: it ensures that they oscillate and that their scalar product with continuous-time polynomials up to degree $M-1$ vanishes. Daubechies wavelets and Coiflets having at least two vanishing moments satisfy these conditions.

Viewing the wavelet $\psi(t)$ as a basic template, define the family $\{\psi_{j,k}, j \in \mathbb{Z}, k \in \mathbb{Z}\}$ of translated and dilated functions

$$\psi_{j,k}(t) = 2^{-j/2} \psi(2^{-j}t - k), \quad j \in \mathbb{Z}, k \in \mathbb{Z}. \tag{2}$$

Positive values of $k$ translate $\psi$ to the right, negative values to the left. The *scale index* $j$ dilates $\psi$ so that large values of $j$ correspond to coarse scales and hence to low frequencies.

Assumptions 1-4 are standard in the context of a multiresolution analysis (MRA) in which case, $\phi$ is the scaling function and $\psi$ is the associated wavelet, see for instance [17, 8]. Daubechies wavelets and Coiflets are examples of orthogonal wavelets constructed using an MRA. In this paper, we do not assume the wavelets to be orthonormal nor that they are associated to a multiresolution analysis. We may therefore work with other convenient choices for $\phi$ and $\psi$ as long as 1-4 are satisfied.

**Discrete Wavelet Transform (DWT) in discrete time.** We now describe how the wavelet coefficients are defined in discrete time, that is for a real-valued sequence $\{x_k, k \in \mathbb{Z}\}$ and for a finite sample $\{x_k, k = 1, \ldots, n\}$. Using the scaling function $\phi$, we first interpolate these discrete values to construct the following continuous-time functions

$$\mathbf{x}_n(t) \stackrel{\text{def}}{=} \sum_{k=1}^{n} x_k \phi(t-k) \quad \text{and} \quad \mathbf{x}(t) \stackrel{\text{def}}{=} \sum_{k \in \mathbb{Z}} x_k \phi(t-k), \quad t \in \mathbb{R}. \tag{3}$$

Without loss of generality we may suppose that the support of the scaling function $\phi$ is included in $[-\Lambda, 0]$ for some integer $\Lambda \geq 1$. Then

$$\mathbf{x}_n(t) = \mathbf{x}(t) \quad \text{for all} \quad t \in [0, n - \Lambda + 1].$$

We may also suppose that the support of the wavelet function $\psi$ is included in $[0, \Lambda]$. With these conventions, the support of $\psi_{j,k}$ is included in the interval $[2^j k, 2^j (k + \Lambda)]$. Let $\tau_0$ be an arbitrary shift order. The wavelet coefficient $W_{j,k}^{\mathbf{x}}$ at scale $j \geq 0$ and location $k \in \mathbb{Z}$ is formally defined as the scalar product in $\mathrm{L}^2(\mathbb{R})$ of the function $t \mapsto \mathbf{x}(t)$ and the wavelet $t \mapsto \psi_{j,k}(t)$:

$$W_{j,k}^{\mathbf{x}} \stackrel{\text{def}}{=} \int_{-\infty}^{\infty} \mathbf{x}(t) \psi_{j,k}(t) \, dt = \int_{-\infty}^{\infty} \mathbf{x}_n(t) \psi_{j,k}(t) \, dt, \quad j \geq 0, k \in \mathbb{Z}, \tag{4}$$

when $[2^j k, 2^j k + \Lambda)] \subseteq [0, n - \Lambda + 1]$, that is, for all $(j,k) \in \mathscr{I}_n$, where $\mathscr{I}_n = \{(j,k) : j \geq 0, 0 \leq k < n_j\}$ with $n_j = 2^{-j}(n - \Lambda + 1) - \Lambda + 1$. It is important to observe that the definition of the wavelet coefficient $W_{j,k}$ at a given index $(j,k)$ does not depend on the sample size $n$ (this is in sharp contrast with Fourier coefficients). For ease of presentation, we will use the convention that at each scale $j$, the first available wavelet coefficient $W_{j,k}$ is indexed by $k = 0$, that is,

$$\mathscr{I}_n \stackrel{\text{def}}{=} \{(j,k) : j \geq 0, 1 \leq k \leq n_j\} \quad \text{with} \quad n_j = 2^{-j}(n - \Lambda + 1) - \Lambda + 1. \tag{5}$$

**Practical implementation.** In practice the DWT of $\{x_k, k = 1, \ldots, n\}$ is not computed using (4) but by linear filtering and decimation. Indeed the wavelet coefficient $W_{j,k}^{\mathbf{x}}$ can be expressed as

$$W_{j,k}^{\mathbf{x}} = \sum_{l \in \mathbb{Z}} x_l \, h_{j,2^j k - l}, \quad (j,k) \in \mathscr{I}_n;, \tag{6}$$

where

$$h_{j,l} \stackrel{\text{def}}{=} 2^{-j/2} \int_{-\infty}^{\infty} \phi(t+l) \psi(2^{-j} t) \, dt. \tag{7}$$

For all $j \geq 0$, the discrete Fourier transform of the transfer function $\{h_{j,l}\}_{l \in \mathbb{Z}}$ is

$$H_j(\lambda) \stackrel{\text{def}}{=} \sum_{l \in \mathbb{Z}} h_{j,l} e^{-i\lambda l} = 2^{-j/2} \int_{-\infty}^{\infty} \sum_{l \in \mathbb{Z}} \phi(t+l) e^{-i\lambda l} \psi(2^{-j} t) \, dt. \tag{8}$$

Since $\phi$ and $\psi$ have compact support, the sum in (8) has only a finite number of non-vanishing terms and, $H_j(\lambda)$ is the transfer function of a finite impulse response filter,

$$H_j(\lambda) = \sum_{l=-\Lambda(2^j+1)+1}^{-1} h_{j,l} e^{-i\lambda l} . \tag{9}$$

When $\phi$ and $\psi$ are the scaling and the wavelet functions associated to a MRA, the wavelet coefficients may be obtained recursively by applying a finite order filter and downsampling by an order 2. This recursive procedure is referred to as *the pyramidal algorithm*, see for instance [17].

**The wavelet spectrum and the scalogram.** Let $X = \{X_t, \ t \in \mathbb{Z}\}$ be a real-valued process with wavelet coefficients $\{W_{j,k}, k \in \mathbb{Z}\}$ and define

$$\sigma_{j,k}^2 = \text{Var}(W_{j,k}) .$$

If $\Delta^M X$ is stationary, by Eq (16) in [19], we have that, for all $j$, the process of its wavelet coefficients at scale $j$, $\{W_{j,k}, k \in \mathbb{Z}\}$, is also stationary. Then, the wavelet variance $\sigma_{j,k}^2$ does not depend on $k$, $\sigma_{j,k}^2 = \sigma_j^2$. The sequence $(\sigma_j^2)_{j \geq 0}$ is called the *wavelet spectrum* of the process $X$.

If moreover $\Delta^M X$ is centered, the wavelet spectrum can be estimated by using the scalogram, defined as the empirical mean of the squared wavelet coefficients computed from the sample $X_1, \ldots, X_n$:

$$\widehat{\sigma}_j^2 = \frac{1}{n_j} \sum_{k=1}^{n_j} W_{j,k}^2 .$$

By [19, Proposition 1], if $K \leq M$, then the scalogram of $X$ can be expressed using the generalized spectral density $f$ appearing in (1) and the filters $H_j$ defining the DWT in (8) as follows:

$$\sigma_j^2 = \int_{-\pi}^{\pi} |H_j(\lambda)|^2 f(\lambda) \, d\lambda, \quad j \geq 0 . \tag{10}$$

# 3 Asymptotic distribution of the W2-CUSUM statistics

## 3.1 The single-scale case

To start with simple presentation and statement of results, we first focus in this section on a test procedure aimed at detecting a change in the variance of the wavelet coefficients at a single scale $j$. Let $X_1, \ldots, X_n$ be the $n$ observations of a time series, and denote by $W_{j,k}$ for $(j,k) \in \mathscr{I}_n$ with $\mathscr{I}_n$ defined in (5) the associated wavelet coefficients. In view of (10), if $X_1, \ldots, X_n$ are a $n$ successive observations of a $K$-th order difference stationary process, then the wavelet variance at each given scale $j$

should be constant. If the process $X$ is not $K$-th order stationary, then it can be expected that the wavelet variance will change either gradually or abruptly (if there is a shock in the original time-series). This thus suggests to investigate the consistency of the variance of the wavelet coefficients.

There are many works aimed at detecting the change point in the variance of a sequence of independent random variables; such problem has also been considered, but much less frequently, for sequences of dependent variables. Here, under the null assumption of $K$-th order difference stationarity, the wavelet coefficients $\{W_{j,k}, k \in \mathbb{Z}\}$ is a covariance stationary sequence whose spectral density is given by (see [19, Corollary 1])

$$\mathbf{D}_{j,0}(\lambda;f) \overset{\text{def}}{=} \sum_{l=0}^{2^j-1} f(2^{-j}(\lambda+2l\pi))2^{-j}\left|H_j(2^{-j}(\lambda+2l\pi))\right|^2 . \tag{11}$$

We will adapt the approach developed in [14], which uses cumulative sum (CUSUM) of squares to detect change points in the variance.

In order to define the test statistic, we first introduce a change point estimator for the mean of the square of the wavelet coefficients at each scale $j$.

$$\widehat{k}_j = \underset{1 \le k \le n_j}{\text{argmax}} \left| \sum_{1 \le i \le k} W_{j,i}^2 - \frac{k}{n_j} \sum_{1 \le i \le n_j} W_{j,i}^2 \right| . \tag{12}$$

Using this change point estimator, the W2-CUSUM statistics is defined as

$$T_{n_j} = \frac{1}{n_j^{1/2} s_{j,n_j}} \left| \sum_{1 \le i \le \widehat{k}_j} W_{j,i}^2 - \frac{\widehat{k}_j}{n_j} \sum_{1 \le i \le n_j} W_{j,i}^2 \right| , \tag{13}$$

where $s_{j,n_j}^2$ is a suitable estimator of the variance of the sample mean of the $W_{j,i}^2$. Because wavelet coefficients at a given scale are correlated, we use the Bartlett estimator of the variance, which is defined by

$$s_{j,n_j}^2 = \widehat{\gamma}_j(0) + 2 \sum_{1 \le l \le q(n_j)} w_l\big(q(n_j)\big)\widehat{\gamma}_j(l) , \tag{14}$$

where

$$\widehat{\gamma}_j(l) \overset{\text{def}}{=} \frac{1}{n_j} \sum_{1 \le i \le n_j - l} (W_{j,i}^2 - \widehat{\sigma}_j^2)(W_{j,i+l}^2 - \widehat{\sigma}_j^2), \tag{15}$$

are the sample autocovariance of $\{W_{j,i}^2, i = 1, \dots, n_j\}$, $\widehat{\sigma}_j^2$ is the scalogram and, for a given integer $q$,

$$w_l(q) = 1 - \frac{l}{1+q}, l \in \{0, \dots, q\} \tag{16}$$

are the so-called Bartlett weights.

The test differs from statistics proposed in [14] only in its denominator, which is the square root of a consistent estimator of the partial sum's variance. If $\{X_n\}$ is short-range dependent, the variance of the partial sum of the scalograms is not simply the sum of the variances of the individual square wavelet coefficient, but also includes the autocovariances of these terms. Therefore, the estimator of the averaged scalogram variance involves not only sums of squared deviations of the scalogram coefficients, but also its weighted autocovariances up to lag $q(n_j)$. The weights $\{w_l(q(n_j))\}$ are those suggested by [21] and always yield a positive sequence of autocovariance, and a positive estimator of the (unnormalized) wavelet spectrum at scale $j$, at frequency zero using a Bartlett window. We will first established the consistency of the estimator $s^2_{j,n_j}$ of the variance of the scalogram at scale $j$ and the convergence of the empirical process of the square wavelet coefficients to the Brownian motion. Denote by $D([0,1])$ is the Skorokhod space of functions which are right continuous at each point of $[0,1)$ with left limit of $(0,1]$ (or *cadlag* functions). This space is, in the sequel, equipped with the classical Skorokhod metric.

**Theorem 3.1.** *Suppose that $X$ is a Gaussian process with generalized spectral density $f$. Let $(\phi, \psi)$ be a scaling and a wavelet function satisfying 1-4. Let $\{q(n_j)\}$ be a non decreasing sequence of integers satisfying*

$$q(n_j) \to \infty \quad and \quad q(n_j)/n_j \to 0 \quad as \quad n_j \to \infty. \tag{17}$$

*Assume that $\Delta^M X$ is non-deterministic and centered, and that $\lambda^{2M} f(\lambda)$ is two times differentiable in $\lambda$ with bounded second order derivative. Then for any fixed scale $j$, as $n \to \infty$,*

$$s^2_{j,n_j} \xrightarrow{P} \frac{1}{\pi} \int_{-\pi}^{\pi} |\mathbf{D}_{j,0}(\lambda;f)|^2 d\lambda, \tag{18}$$

*where $\mathbf{D}_{j,0}(\lambda;f)$ is the wavelet coefficients spectral density at scale $j$ see (11). Moreover, defining $\sigma_j^2$ by (10),*

$$\frac{1}{n_j^{1/2} s_{j,n_j}} \sum_{i=1}^{[n_j t]} \left( W_{j,i}^2 - \sigma_j^2 \right) \xrightarrow{\mathscr{L}} B(t) \quad in \quad D([0,1]), \quad as \quad n \to \infty \tag{19}$$

*where $(B(t), t \in [0,1])$ is the standard Brownian motion.*

*Remark 0.1.* The fact that $X$ is Gaussian can be replaced by the more general assumption that the process $X$ is linear in the strong sense, under appropriate moment conditions on the innovation. The proofs are then more involved, especially to establish the invariance principle which is pivotal in our derivation.

*Remark 0.2.* By allowing $q(n_j)$ to increase but at a slower rate than the number of observations, the estimator of the averaged scalogram variance adjusts appropriately for general forms of short-range dependence among the scalogram coefficients. Of course, although the condition (17) ensure the consistency of $s^2_{j,n_j}$, they provide little guidance in selecting a truncation lag $q(n_j)$. When $q(n_j)$ becomes large relative to the sample size $n_j$, the finite-sample distribution of the test statistic might be far from its asymptotic limit. However $q(n_j)$ cannot be chosen too small since the autocovariances beyond lag $q(n_j)$ may be significant and should be included in the weighted sum. Therefore, the truncation lag must be chosen ideally using some data-driven procedures. [1] and [22] provide a data-dependent rule for choosing $q(n_j)$. These contributions suggest that selection of bandwidth according to an asymptotically optimal procedure tends to lead to more accurately sized test statistics than do traditional procedure The methods suggested by [1] for selecting the bandwidth optimally is a plug-in approach. This procedure require the researcher to fit an ARMA model of given order to provide a rough estimator of the spectral density and of its derivatives at zero frequencies (although misspecification of the order affects only optimality but not consistency). The minimax optimality of this method is based on an asymptotic mean-squared error criterion and its behavior in the finite sample case is not precisely known. The procedure outlined in [22] suggests to bypass the modeling step, by using instead a pilot truncated kernel estimates of the spectral density and its derivative. We use these data driven procedures in the Monte Carlo experiments (these procedures have been implemented in the *R-package sandwich.*

*Proof.* Since $X$ is Gaussian and $\Delta^M X$ is centered, Eq. (17) in [19] implies that $\{W_{j,k}, k \in \mathbb{Z}\}$ is a centered Gaussian process, whose distribution is determined by

$$\gamma_j(h) = \text{cov}(W_{j,0}, W_{j,h}) = \int_{-\pi}^{\pi} \mathbf{D}_{j,0}(\lambda; f) e^{-i\lambda h} d\lambda .$$

From Corollary 1 and equation (16) in [19], we have

$$\mathbf{D}_{j,0}(\lambda; f)$$

$$= \sum_{l=0}^{2^j-1} f\left(2^{-j}(\lambda + 2l\pi)\right) 2^{-j} \left|\widetilde{H}_j(2^{-j}(\lambda + 2l\pi))\right|^2 \left|1 - e^{-i2^{-j}(\lambda + 2l\pi)}\right|^{2M},$$

where $\widetilde{H}_j$ is a trigonometric polynomial. Using that

$$|1 - e^{-i\xi}|^{2M} = |\xi|^{2M} \left|\frac{1 - e^{-i\xi}}{i\xi}\right|^{2M}$$

and that $|\xi|^{2M} f(\xi)$ has a bounded second order derivative, we get that $\mathbf{D}_{j,0}(\lambda; f)$ has also a bounded second order derivative. In particular,

$$\int_{-\pi}^{\pi} |\mathbf{D}_{j,0}(\lambda; f)|^2 d\lambda < \infty \quad \text{and} \quad \sum_{s \in \mathbb{Z}} |\gamma_j(s)| < \infty . \tag{20}$$

The proof may be decomposed into 3 steps. We first prove the consistency of the Bartlett estimator of the variance of the squares of wavelet coefficients $s_{j,n_j}^2$, that is (18). Then we determine the asymptotic normality of the finite-dimensional distributions of the empirical scalogram, suitably centered and normalized. Finally a tightness criterion is proved, to establish the convergence in the Skorokhod space. Combining these three steps completes the proof of (19).

**Step 1**. Observe that, by the Gaussian property, $\mathrm{cov}(W_{j,0}^2, W_{j,h}^2) = 2\gamma_j^2(h)$. Using Theorem 3-i in [10], the limit (18) follows from

$$2 \sum_{h=-\infty}^{+\infty} \gamma_j^2(h) = \frac{1}{\pi} \int_{-\pi}^{\pi} |\mathbf{D}_{j,0}(\lambda;f)|^2 d\lambda < \infty, \tag{21}$$

and

$$\sup_{h \in \mathbb{Z}} \sum_{r,s=-\infty}^{+\infty} |\mathscr{K}(h,r,s)| < \infty. \tag{22}$$

where

$$\mathscr{K}(h,r,s) = \mathrm{Cum}\left(W_{j,k}^2, W_{j,k+h}^2, W_{j,k+r}^2, W_{j,k+s}^2\right). \tag{23}$$

Equation (21) follows from Parseval's equality and (20). Let us now prove (22). Using that the wavelet coefficients are Gaussian, we obtain

$$\mathscr{K}(h,r,s) = 12\{\gamma_j(h)\gamma_j(r-s)\gamma_j(h-r)\gamma_j(s)$$
$$+ \gamma_j(h)\gamma_j(r-s)\gamma_j(h-s)\gamma_j(r) + \gamma_j(s-h)\gamma_j(r-h)\gamma_j(r)\gamma_j(s)\}.$$

The bound of the last term is given by

$$\sup_{h \in \mathbb{Z}} \sum_{r,s=-\infty}^{+\infty} |\gamma_j(s-h)\gamma_j(r-h)\gamma_j(r)\gamma_j(s)| \le \sup_{h} \left( \sum_{r=-\infty}^{+\infty} |\gamma_j(r)\gamma_j(r-h)| \right)^2$$

which is finite by the Cauchy-Schwarz inequality, since $\sum_{r \in \mathbb{Z}} \gamma_j^2(r) < \infty$.

Using $|\gamma_j(h)| < \gamma_j(0)$ and the Cauchy-Schwarz inequality, we have

$$\sup_{h \in \mathbb{Z}} \sum_{r,s=-\infty}^{+\infty} |\gamma_j(h)\gamma_j(r-s)\gamma_j(h-r)\gamma_j(s)| \le \gamma_j(0) \sum_{u \in \mathbb{Z}} \gamma_j^2(u) \sum_{s \in \mathbb{Z}} |\gamma_j(s)|,$$

and the same bound applies to

$$\sup_{h \in \mathbb{Z}} \sum_{r,s=-\infty}^{+\infty} |\gamma_j(h)\gamma_j(r-s)\gamma_j(h-s)\gamma_j(r)|.$$

Hence,we have (22) by (20), which achieves the proof of Step 1.
**Step 2**. Let us define

$$S_{n_j}(t) = \frac{1}{\sqrt{n_j}} \sum_{i=1}^{\lfloor n_j t \rfloor} (W_{j,i}^2 - \sigma_j^2), \tag{24}$$

where $\sigma_j^2 = \mathbb{E}(W_{j,i}^2)$, and $\lfloor x \rfloor$ is the entire part of $x$. Step 2 consists in proving that for $0 \le t_1 \le \ldots \le t_k \le 1$, and $\mu_1, \ldots, \mu_k \in \mathbb{R}$,

$$\sum_{i=1}^{k} \mu_i S_{n_j}(t_i) \xrightarrow{\mathscr{L}} \mathscr{N}\left(0, \frac{1}{\pi}\int_{-\pi}^{\pi} \left|\mathbf{D}_{j,0}(\lambda;f)\right|^2 d\lambda \times \mathrm{Var}\left(\sum_{i=1}^{k} \mu_i B(t_i)\right)\right). \quad (25)$$

Observe that

$$\sum_{i=1}^{k} \mu_i S_{n_j}(t_i) = \frac{1}{\sqrt{n_j}}\sum_{i=1}^{k} \mu_i \sum_{l=1}^{n_j} (W_{j,l}^2 - \sigma_j^2)\mathbb{1}_{\{l\le\lfloor n_j t_i\rfloor\}}$$

$$= \sum_{l=1}^{n_j} W_{j,l}^2 a_{l,n} - E\left(\sum_{l=1}^{n_j} W_{j,l}^2 a_{l,n}\right)$$

$$= \xi_{n_j}^T A_{n_j} \xi_{n_j},$$

where we set $a_{l,n} = \frac{1}{\sqrt{n_j}}\sum_{i=1}^{k} \mu_i \mathbb{1}_{\{l\le\lfloor n_j t_i\rfloor\}}$, $\xi_{n_j} = (W_{j1},\ldots,W_{jn_j})^T$ and $A_{n_j}$ is the diagonal matrix with diagonal entries $(a_{1,n_j},\ldots,a_{n_j,n_j})$. Applying [20, Lemma 12], (25) is obtained by proving that, as $n_j \to \infty$,

$$\rho(A_{n_j})\rho(\Gamma_{n_j}) \to 0 \quad (26)$$

$$\mathrm{Var}\left(\sum_{i=1}^{k} \mu_i S_{n_j}(t_i)\right) \to \frac{1}{\pi}\int_{-\pi}^{\pi} \left|\mathbf{D}_{j,0}(\lambda;f)\right|^2 d\lambda \times \mathrm{Var}\left(\sum_{i=1}^{k} \mu_i(B(t_i))\right), \quad (27)$$

where $\rho(A)$ denote the spectral radius of the matrix $A$, that is, the maximum modulus of its eigenvalues and $\Gamma_{n_j}$ is the covariance matrix of $\xi_{n_j}$. The process $(W_{j,i})_{\{i=1,\ldots,n_j\}}$ is stationary with spectral density $\mathbf{D}_{j,0}(.;f)$. Thus, by Lemma 2 in [19] its covariance matrix $\Gamma_{n_j}$ satisfies $\rho(\Gamma_{n_j}) \le 2\pi\sup_{\lambda}\mathbf{D}_{j,0}(\lambda;f)$. Furthermore, as $n_j \to \infty$,

$$\rho(A_{n_j}) = \max_{1\le l\le n_j} \frac{1}{\sqrt{n_j}}\left|\sum_{i=1}^{k} \mu_i \mathbb{1}_{\{l\le\lfloor n_j t_i\rfloor\}}\right| \le n_j^{-1/2}\sum_{i=1}^{k} |\mu_i| \to 0,$$

and (26) holds. We now prove (27). Using that $B(t)$ has variance $t$ and independent and stationary increments, and that these properties characterize its covariance function, it is sufficient to show that, for all $t \in [0,1]$, as $n_j \to \infty$,

$$\mathrm{Var}\left(S_{n_j}(t)\right) \to t\int_{-\pi}^{\pi} \left|\mathbf{D}_{j,0}(\lambda;f)\right|^2 d\lambda, \quad (28)$$

and for all $0 \le r \le s \le t \le 1$, as $n_j \to \infty$,

$$\mathrm{cov}\left(S_{n,j}(t) - S_{n,j}(s), S_{n,j}(r)\right) \to 0. \quad (29)$$

For any sets $A, B \subseteq [0,1]$, we set

$$V_{n_j}(\tau,A,B) = \frac{1}{n_j}\sum_{k\ge 1} \mathbb{1}_A((k+\tau)/n_j)\mathbb{1}_B(k/n_j).$$

For all $0 \leq s,t \leq 1$, we have

$$\text{cov}\left(S_{n_j}(t), S_{n_j}(s)\right) = \frac{1}{n_j} \sum_{i=1}^{\lfloor n_j t \rfloor} \sum_{k=1}^{\lfloor n_j s \rfloor} \text{cov}(W_{j,i}^2, W_{j,k}^2)$$

$$= 2 \sum_{\tau \in \mathbb{Z}} \gamma_j^2(\tau) V_{n_j}(\tau, ]0,t], ]0,s]) \,.$$

The previous display applies to the left-hand side of (28) when $s = t$ and for $0 \leq r \leq s \leq t \leq 1$, it yields

$$\text{cov}\left(S_{n_j}(t) - S_{n_j}(s), S_{n_j}(r)\right) = 2 \sum_{\tau \in \mathbb{Z}} \gamma_j^2(\tau) V_{n_j}(\tau, ]s,t], ]0,r]) \,.$$

Observe that for all $A, B \subseteq [0,1]$, $\sup_\tau |V_n(j, \tau, A, B| \leq \frac{k}{n_j} \leq 1$. Hence, by dominated convergence, the limits in (28) and (29) are obtained by computing the limits of $V_n(j, \tau, ]0,t], ]0,t])$ and $V_n(j, \tau, ]s,t], ]0,r])$ respectively. We have for any $\tau \in \mathbb{Z}, t > 0$, and $n_j$ large enough,

$$\sum_{k \geq 1} \mathbb{1}_{\{\frac{k+\tau}{n_j} \in ]0,t]\}} \mathbb{1}_{\{\frac{k}{n_j} \in ]0,t]\}} = \left\{(n_j t \wedge n_j t - \tau)\right\}_+ = n_j t - \tau_+ \,.$$

Hence, as $n_j \to \infty$, $V_{n_j}(\tau, ]0,t], ]0,t]) \to t$ and, by (21), (28) follows. We have for any $\tau \in \mathbb{Z}$ and $0 < r \leq s \leq t$,

$$\sum_{k \geq 1} \mathbb{1}_{\{\frac{k+\tau}{n_j} \in ]s,t]\}} \mathbb{1}_{\{\frac{k}{n_j} \in ]0,r]\}} = \left\{(n_j r \wedge \{n_j t - \tau\}) - (0 \vee \{n_j s - \tau\})\right\}_+$$

$$= (n_j r - n_j s + \tau)_+ \to \mathbb{1}_{\{r=s\}} \tau_+ \,,$$

where the last equality holds for $n_j$ large enough and the limit as $n_j \to \infty$. Hence $V_{n_j}(\tau, ]s,t], ]0,r]) \to 0$ and (29) follows, which achieves Step 2.

**Step 3.** We now prove the tightness of $\{S_{n_j}(t), t \in [0,1]\}$ in the Skorokhod metric space. By Theorem 13.5 in [4], it is sufficient to prove that for all $0 \leq r \leq s \leq t$,

$$\mathbb{E}\left[|S_{n_j}(s) - S_{n_j}(r)|^2 |S_{n_j}(t) - S_{n_j}(s)|^2\right] \leq C|t - r|^2 \,,$$

where $C > 0$ is some constant independent of $r, s, t$ and $n_j$. We shall prove that, for all $0 \leq r \leq t$,

$$\mathbb{E}\left[|S_{n_j}(t) - S_{n_j}(r)|^4\right] \leq C_1 \{n_j^{-1}(\lfloor n_j t \rfloor - \lfloor n_j r \rfloor)\}^2 \,. \tag{30}$$

By the Cauchy-Schwarz inequality, and using that, for $0 \leq r \leq s \leq t$,

$$n_j^{-1}(\lfloor n_j t \rfloor - \lfloor n_j s \rfloor) \times n_j^{-1}(\lfloor n_j s \rfloor - \lfloor n_j r \rfloor) \leq 4(t - r)^2 \,,$$

the criterion (30) implies the previous criterion. Hence the tightness follows from (30), that we now prove. We have, for any $\mathbf{i} = (i_1, \ldots, i_4)$,

$$\mathbb{E}\left[\prod_{k=1}^{4}(W_{j,i_k}^2 - \sigma_j^2)\right] = \mathrm{Cum}(W_{j,i_1}^2,\ldots,W_{j,i_4}^2) + \mathbb{E}(W_{j,i_1}^2,W_{j,i_2}^2)\mathbb{E}[W_{j,i_3}^2,W_{j,i_4}^2]$$
$$+ \mathbb{E}[W_{j,i_1}^2,W_{j,i_3}^2]\mathbb{E}[W_{j,i_2}^2 W_{i_4}^2] + \mathbb{E}[W_{j,i_1}^2,W_{i_4}^2]\mathbb{E}[W_{j,i_2}^2 W_{i_3}^2] \ .$$

It follows that, denoting for $0 \le r \le t \le 1$,

$$\mathbb{E}\left[|S_{n_j}(t) - S_{n_j}(r)|^4\right] = \frac{1}{n_j^2}\sum_{i \in A_{r,t}^4} \mathrm{Cum}(W_{j,i_1}^2,\ldots,W_{j,i_4}^2)$$

$$+ \frac{3}{n_j^2}\left(\sum_{i \in A_{r,t}^2}\mathbb{E}[W_{j,i_1} W_{j,i_2}]\right)^2$$

where $A_{r,t} = \{\lfloor n_j r\rfloor + 1,\ldots,\lfloor n_j t\rfloor\}$. Observe that

$$0 \le \frac{1}{n_j}\sum_{i \in A_{r,t}^2}\mathbb{E}[W_{j,i_1}^2 W_{j,i_2}^2] \le 2\sum_{\tau \in \mathbb{Z}}\gamma_j^2(\tau) \times n_j^{-1}(\lfloor n_j t\rfloor - \lfloor n_j r\rfloor) \ .$$

Using that, by (23), $\mathrm{Cum}(W_{j,i_1}^2,\ldots,W_{j,i_4}^2) = \mathscr{K}(i_2 - i_1, i_3 - i_1, i_4 - i_1)$, we have

$$\sum_{i \in A_{r,t}^4}|\mathrm{Cum}(W_{j,i_1}^2,\ldots,W_{j,i_4}^2)| \le (\lfloor n_j t\rfloor - \lfloor n_j r\rfloor)\sum_{h,s,l=\lfloor n_j r\rfloor - \lfloor n_j t\rfloor + 1}^{\lfloor n_j t\rfloor - \lfloor n_j r\rfloor - 1}|\mathscr{K}(h,s,l)|$$

$$\le 2(\lfloor n_j t\rfloor - \lfloor n_j r\rfloor)^2 \sup_{h \in \mathbb{Z}}\sum_{r,s=-\infty}^{+\infty}|\mathscr{K}(h,r,s)| \ .$$

The last three displays and (22) imply (30), which proves the tightness.

Finally, observing that the variance (21) is positive, unless $f$ vanishes almost everywhere, the convergence (19) follows from Slutsky's lemma and the three previous steps.

## 3.2 The multiple-scale case

The results above can be extended to test simultaneously changes in wavelet variances occurring simultaneously at multiple time-scales. To construct a multiple scale test, consider the *between-scale* process

$$\{[W_{j,k}^X, \mathbf{W}_{j,k}^X(j-j')^T]^T\}_{k \in \mathbb{Z}} \ , \tag{31}$$

where the superscript $^T$ denotes the transpose and $\mathbf{W}_{j,k}^X(u)$, $u = 0, 1,\ldots,j$, is defined as follows:

$$\mathbf{W}_{j,k}^X(u) \stackrel{\text{def}}{=} \left[W_{j-u,2^uk}^X, W_{j-u,2^uk+1}^X, \ldots, W_{j-u,2^uk+2^u}^X \; 1\right]^T . \tag{32}$$

It is a $2^u$-dimensional vector of wavelet coefficients at scale $j' = j - u$ and involves all possible translations of the position index $2^u k$ by $v = 0, 1, \ldots, 2^u - 1$. The index $u$ in (32) denotes the scale difference $j - j' \geq 0$ between the finest scale $j'$ and the coarsest scale $j$. Observe that $\mathbf{W}_{j,k}^X(0)$ ($u = 0$) is the scalar $W_{j,k}^X$. It is shown in [19, Corollary 1] that, when $\Delta^M X$ is covariance stationary, the between scale process $\{[W_{j,k}^X, \mathbf{W}_{j,k}^X(j-j')^T]^T\}_{k \in \mathbb{Z}}$ is also covariance stationary. Moreover, for all $0 \leq u \leq j$, the *between scale covariance matrix* is defined as

$$\text{Cov}\left(W_{j,0}^X, \mathbf{W}_{j,k}^X(u)\right) = \int_{-\pi}^{\pi} e^{i\lambda k} \mathbf{D}_{j,u}(\lambda;f) \, d\lambda , \tag{33}$$

where $\mathbf{D}_{j,u}(\lambda;f)$ is the cross-spectral density function of the between-scale process given by (see [19, Corollary 1])

$$\mathbf{D}_{j,u}(\lambda;f) \stackrel{\text{def}}{=} \sum_{l=0}^{2^j-1} \mathbf{e}_u(\lambda + 2l\pi) f(2^{-j}(\lambda + 2l\pi)) 2^{-j/2} H_j(2^{-j}(\lambda + 2l\pi))$$

$$\times 2^{-(j-u)/2} \overline{H_{j-u}(2^{-j}(\lambda + 2l\pi))} , \tag{34}$$

where for all $\xi \in \mathbb{R}$,

$$\mathbf{e}_u(\xi) \stackrel{\text{def}}{=} 2^{-u/2} \left[1, e^{-i2^{-u}\xi}, \ldots, e^{-i(2^u-1)2^{-u}\xi}\right]^T .$$

The case $u = 0$ corresponds to the spectral density of the *within-scale* process $\{W_{j,k}\}_{k \in \mathbb{Z}}$ given in (11). Under the null hypothesis that $X$ is $K$-th order stationary, a *multiple scale* procedure aims at testing that the scalogram in a range satisfies

$$\mathcal{H}_0 : \sigma_{j,1}^2 = \cdots = \sigma_{j,n_j}^2, \text{for all } j \in \{J_1, J_1+1, \ldots, J_2\} \tag{35}$$

where $J_1$ and $J_2$ are the *finest* and the *coarsest* scales included in the procedure, respectively. The wavelet coefficients at different scales are not uncorrelated so that both the *within-scale* and the *between scale* covariances need to be taken into account.

As before, we use a CUSUM statistic in the wavelet domain. However, we now use multiple scale vector statistics. Consider the following process

$$Y_{J_1,J_2,i} = \left(W_{J_2,i}^2, \sum_{u=1}^{2} W_{J_2-1,2(i-1)+u}^2, \ldots, \sum_{u=1}^{2^{(J_2-J_1)}} W_{J_1,2^{(J_2-J_1)}(i-1)+u}^2\right)^T .$$

The *Bartlett estimator* of the covariance matrix of the square wavelet's coefficients for scales $\{J_1, \ldots, J_2\}$ is the $(J_2 - J_1 + 1) \times (J_2 - J_1 + 1)$ symmetric definite positive matrix $\widehat{\Gamma}_{J_1,J_2}$ given by :

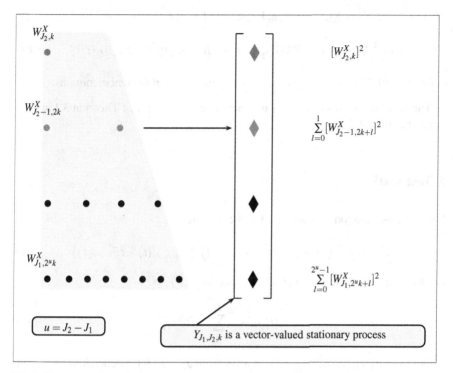

**Fig. 1** Between scale stationary process.

$$\widehat{\Gamma}_{J_1,J_2} = \sum_{\tau=-q(n_{J_2})}^{q(n_{J_2})} w_\tau [q(n_{J_2})] \widehat{\gamma}_{J_1,J_2}(\tau) , \text{ where} \tag{36}$$

$$\widehat{\gamma}_{J_1,J_2}(\tau) = \frac{1}{n_{J_2}} \sum_{i,i+\tau=1}^{n_{J_2}} (Y_{J_1,J_2,i} - \bar{Y}_{J_1,J_2})(Y_{J_1,J_2,i+\tau} - \bar{Y}_{J_1,J_2})^T . \tag{37}$$

where $\bar{Y}_{J_1,J_2} = \frac{1}{n_{J_2}} \sum_{i=1}^{n_{J_2}} Y_{J_1,J_2,i}$

Finally, let us define the vector of partial sum from scale $J_1$ to scale $J_2$ as

$$S_{J_1,J_2}(t) = \frac{1}{\sqrt{n_{J_2}}} \left[ \sum_{i=1}^{\lfloor n_j t \rfloor} W_{j,i}^2 \right]_{j=J_1,\ldots,J_2} . \tag{38}$$

**Theorem 3.2.** *Under the assumptions of Theorem 3.1, we have, as $n \to \infty$,*

$$\widehat{\Gamma}_{J_1,J_2} = \Gamma_{J_1,J_2} + O_P\left(\frac{q(n_{J_2})}{n_{J_2}}\right) + O_P(q^{-1}(n_{J_2})), \tag{39}$$

where $\Gamma_{J_1,J_2}(j,j') = \sum_{h\in\mathbb{Z}} \mathrm{cov}(Y_{j,0}, Y_{j',h})$, with $1 \leq j, j' \leq J_2 - J_1 + 1$ and,

$$\hat{\Gamma}_{J_1,J_2}^{-1/2} \left(S_{J_1,J_2}(t) - \mathbb{E}[S_{J_1,J_2}(t)]\right) \xrightarrow{\mathcal{L}} B(t) = (B_{J_1}(t),\ldots,B_{J_2}(t)), \tag{40}$$

in $D^{J_2-J_1+1}[0,1]$, where $\{B_j(t)\}_{j=J_1,\ldots,J_2}$ are independent Brownian motions.

The proof of this result follows the same line as the proof of Theorem 3.1 and is therefore omitted.

## 4 Test statistics

Under the assumption of Theorem 3.1, the statistics

$$T_{J_1,J_2}(t) \stackrel{\text{def}}{=} (S_{J_1,J_2}(t) - tS_{J_1,J_2}(1))^T \hat{\Gamma}_{J_1,J_2}^{-1} (S_{J_1,J_2}(t) - tS_{J_1,J_2}(1)) \tag{41}$$

converges in weakly in the Skorokhod space $D([0,1])$

$$T_{J_1,J_2}(t) \xrightarrow{\mathcal{L}} \sum_{\ell=1}^{J_2-J_1-1} \left[B_\ell^0(t)\right]^2 \tag{42}$$

where $t \mapsto (B_1^0(t),\ldots,B_{J_2-J_1+1}^0(t))$ is a vector of $J_2 - J_1 + 1$ independent Brownian bridges

For any continuous function $F : D[0,1] \to \mathbb{R}$, the continuous mapping Theorem implies that

$$F[T_{J_1,J_2}(\cdot)] \xrightarrow{\mathcal{L}} F\left[\sum_{\ell=1}^{J_2-J_1-1} \left[B_\ell^0(\cdot)\right]^2\right].$$

We may for example apply either integral or max functionals, or weighted versions of these. A classical example of integral function is the so-called Cramér-Von Mises functional given by

$$\mathrm{CVM}(J_1,J_2) \stackrel{\text{def}}{=} \int_0^1 T_{J_1,J_2}(t)\mathrm{d}t , \tag{43}$$

which converges to $C(J_2 - J_1 + 1)$ where for any integer $d$,

$$C(d) \stackrel{\text{def}}{=} \int_0^1 \sum_{\ell=1}^{d} \left[B_\ell^0(t)\right]^2 \mathrm{d}t . \tag{44}$$

The test rejects the null hypothesis when $\mathrm{CVM}_{J_1,J_2} \geq c(J_2 - J_1 + 1,\alpha)$, where $c(d,\alpha)$ is the $1 - \alpha$th quantile of the distribution of $C(d)$. The distribution of the random variable $C(d)$ has been derived by [15] (see also [7] for more recent references). It holds that, for $x > 0$,

| Nominal S. | $d=1$ | $d=2$ | $d=3$ | $d=4$ | $d=5$ | $d=6$ |
|---|---|---|---|---|---|---|
| 0.95 | 0.4605 | 0.7488 | 1.0014 | 1.2397 | 1.4691 | 1.6848 |
| 0.99 | 0.7401 | 1.0721 | 1.3521 | 1.6267 | 1.8667 | 2.1259 |

**Table 1** Quantiles of the distribution $C(d)$ (see (44)) for different values of $d$

| $d$ | 1 | 2 | 3 | 4 | 5 | 6 |
|---|---|---|---|---|---|---|
| 0.95 | 1.358 | 1.58379 | 1.7472 | 1.88226 | 2.00 | 2.10597 |
| 0.99 | 1.627624 | 1.842726 | 2.001 | 2.132572 | 2.24798 | 2.35209 |

**Table 2** Quantiles of the distribution $D(d)$ (see (46)) for different values of $d$.

$$\mathbb{P}(C(d) \leq x) = \frac{2^{(d+1)/2}}{\pi^{1/2} x^{d/4}} \sum_{j=0}^{\infty} \frac{\Gamma(j+d/2)}{j! \Gamma(d/2)} e^{-(j+d/4)^2/x} \mathrm{Cyl}_{(d-2)/2} \left( \frac{2j+d/2}{x^{1/2}} \right)$$

where $\Gamma$ denotes the gamma function and Cyl are the parabolic cylinder functions. The quantile of this distribution are given in table 1 for different values of $d = J_2 - J_1 + 1$. It is also possible to use the max. functional leading to an analogue of the Kolmogorov-Smirnov statistics,

$$\mathrm{KSM}(J_1, J_2) \stackrel{\text{def}}{=} \sup_{0 \leq t \leq 1} T_{J_1, J_2}(t) \tag{45}$$

which converges to $D(J_2 - J_1 + 1)$ where for any integer $d$,

$$D(d) \stackrel{\text{def}}{=} \sup_{0 \leq t \leq 1} \sum_{\ell=1}^{d} \left[ B_\ell^0(t) \right]^2 . \tag{46}$$

The test reject the null hypothesis when $\mathrm{KSM}_{J_1, J_2} \geq \delta(J_2 - J_1 + 1, \alpha)$, where $\delta(d, \alpha)$ is the $(1 - \alpha)$-quantile of $D(d)$. The distribution of $D(d)$ has again be derived by [15] (see also [23] for more recent references). It holds that, for $x > 0$,

$$\mathbb{P}(D(d) \leq x) = \frac{2^{1+(2-d)/2}}{\Gamma(d/2) a^d} \sum_{n=1}^{\infty} \frac{j_{v,n}^{2v}}{J_{v+1}^2(j_{v,n})} \exp\left( -\frac{j_{v,n}^2}{2x^2} \right),$$

where $0 < j_{v,1} < j_{v,2} < \dots$ is the sequence of positive zeros of $J_v$, the Bessel function of index $v = (d-2)/2$. The quantiles of this distribution are given in Table 2.

## 5 Power of the W2-CUSUM statistics

### 5.1 Power of the test in single scale case

In this section we investigate the power of the test. A minimal requirement is to establish that the test procedure is pointwise consistent in a presence of a breakpoint, *i.e.* that under a fixed alternative, the probability of detection converges to one as the sample size goes to infinity. We must therefore first define such alternative. For simplicity, we will consider an alternative where the process exhibit a single breakpoint, though it is likely that the test does have power against more general class of alternatives.

The alternative that we consider in this section is defined as follows. Let $f_1$ and $f_2$ be two given generalized spectral densities and suppose that, at a given scale $j$, $\int_{-\pi}^{\pi} |H_j(\lambda)|^2 f_i(\lambda) d\lambda < \infty$, $i = 1, 2$, and

$$\int_{-\pi}^{\pi} |H_j(\lambda)|^2 (f_1(\lambda) - f_2(\lambda)) d\lambda \neq 0. \tag{47}$$

Define by $(X_{l,i})_{l \in \mathbb{Z}}$, $i = 1, 2$, be two Gaussian processes, defined on the same probability space, with generalized spectral density $f_1$. We do not specify the dependence structure between these two processes, which can be arbitrary. Let $\kappa \in ]0, 1[$ be a breakpoint. We consider a sequence of Gaussian processes $(X_k^n)_{k \in \mathbb{Z}}$, such that

$$X_k^{(n)} = X_{k,i} \text{ for } k \leq \lfloor n\kappa \rfloor \text{ and } X_k^{(n)} = X_{k,2} \text{ for } k \geq \lfloor n\kappa \rfloor + 1. \tag{48}$$

**Theorem 5.1.** *Consider $\{X_k^n\}_{k \in \mathbb{Z}}$ be a sequence of processes specified by (47) and (48). Assume that $q(n_j)$ is non decreasing and :*

$$q(n_j) \to \infty \text{ and } \frac{q(n_j)}{n_j} \to 0 \text{ as } n_j \to \infty. \tag{49}$$

*Then the statistic $T_{n_j}$ defined by (13) satisfies*

$$\frac{\sqrt{n_j}}{\sqrt{2q(n_j)}} \sqrt{\kappa(1-\kappa)}(1 + o_p(1)) \leq T_{n_j} \xrightarrow{P} \infty. \tag{50}$$

*Proof.* Let $k_j = \lfloor n_j \kappa \rfloor$ the change point in the wavelet spectrum at scale $j$. We write $q$ for $q(n_j)$ and suppress the dependence in $n$ in this proof to alleviate the notation. By definition $T_{n_j} = \frac{1}{s_{j,n_j}} \sup_{0 \leq t \leq 1} (S_{n_j}(t) - tS_{n_j}(1))$, where the process $t \mapsto S_{n_j}(t)$ is defined in (24). Therefore, $T_{n_j} \geq \frac{1}{s_{j,n_j}} (S_{n_j}(\kappa) - \kappa S_{n_j}(1))$. The proof consists in establishing that $\frac{1}{s_{j,n_j}} (S_{n_j}(\kappa) - \kappa S_{n_j}(1)) = \frac{\sqrt{n_j}}{\sqrt{2q(n_j)}} \sqrt{\kappa(1-\kappa)}(1 + o_p(1))$. We first decompose this difference as follows

$$S_{n_j}(\kappa) - \kappa S_{n_j}(1) = \frac{1}{\sqrt{n_j}} \left| \sum_{i=1}^{\lfloor n_j \kappa \rfloor} W_{j,i}^2 - \kappa \sum_{i=1}^{n_j} W_{j,i}^2 \right|$$

$$= B_{n_j} + f_{n_j}$$

where $B_{n_j}$ is a fluctuation term

$$B_{n_j} = \frac{1}{\sqrt{n_j}} \left| \sum_{i=1}^{k_j} (W_{j,i}^2 - \sigma_{j,i}^2) - \kappa \sum_{i=1}^{n_j} (W_{j,i}^2 - \sigma_{j,i}^2) \right| \tag{51}$$

and $f_{n_j}$ is a bias term

$$f_{n_j} = \frac{1}{\sqrt{n_j}} \left| \sum_{i=1}^{k_j} \sigma_{j,i}^2 - \kappa \sum_{i=1}^{n_j} \sigma_{j,i}^2 \right| . \tag{52}$$

Since support of $h_{j,l}$ is included in $[-\Lambda(2^j+1), 0]$ where $h_{j,l}$ is defined in (7), there exits a constant $a > 0$ such that

$$W_{j,i} = W_{j,i;1} = \sum_{l \leq k} h_{j,2^j i - l} X_{l,1}, \text{ for } i < k_j, \tag{53}$$

$$W_{j,i} = W_{j,i;2} = \sum_{l > k} h_{j,2^j i - l} X_{l,2} \text{ for } i > k_j + a, \tag{54}$$

$$W_{j,i} = \sum_{l} h_{j,2^j i - l} X_l, \text{ for } k_j \leq i < k_j + a. \tag{55}$$

Since the process $\{X_{l,1}\}_{l \in \mathbb{Z}}$ and $\{X_{l,2}\}_{l \in \mathbb{Z}}$ are both $K$-th order covariance stationary, the two processes $\{W_{j,i;1}\}_{i \in \mathbb{Z}}$ and $\{W_{j,i;2}\}_{i \in \mathbb{Z}}$ are also covariance stationary. The wavelet coefficients $W_{j,i}$ for $i \in \{k_j, \ldots, k_j + a\}$ are computed using observations from the two processes $X_1$ and $X_2$. Let us show that there exits a constant $C > 0$ such that, for all integers $l$ and $\tau$,

$$\text{Var} \left( \sum_{i=l}^{l+\tau} W_{j,i}^2 \right) \leq C\tau . \tag{56}$$

Using (21), we have, for $\varepsilon = 1, 2$,

$$\text{Var} \left( \sum_{i=l}^{l+\tau} W_{j,i;\varepsilon}^2 \right) \leq \frac{\tau}{\pi} \int_{-\pi}^{\pi} |D_{j,0;\varepsilon}(\lambda)|^2 \, d\lambda$$

where, $D_{j,0;1}(\lambda)$ and $D_{j,0;2}(\lambda)$ denote the spectral density of the stationary processes $\{W_{j,i;1}\}_{i \in \mathbb{Z}}$ and $\{W_{j,i;2}\}_{i \in \mathbb{Z}}$ respectively. Using Minkowski inequality, we have for $l \leq k_j \leq k_j + a < l + \tau$ that $\left( \text{Var} \sum_{i=l}^{l+\tau} W_{j,i}^2 \right)^{1/2}$ is at most

$$\left(\operatorname{Var}\sum_{i=l}^{k_j} W_{j,i}^2\right)^{1/2} + \sum_{i=k_j+1}^{k_j+a} \left(\operatorname{Var}W_{j,i}^2\right)^{1/2} + \left(\operatorname{Var}\sum_{i=k_j+a+1}^{l+\tau} W_{j,i}^2\right)^{1/2}$$

$$\leq \left(\operatorname{Var}\sum_{i=l}^{k_j} W_{j,i;1}^2\right)^{1/2} + a\sup_i (\operatorname{Var}W_{j,i}^2)^{1/2} + \left(\operatorname{Var}\sum_{i=k_j+a+1}^{l+\tau} W_{j,i;2}^2\right)^{1/2}.$$

Observe that $\operatorname{Var}(W_{j,i}^2) \leq 2(\sum_l |h_{j,l}|)^2 \left(\sigma_{j,1}^2 \vee \sigma_{j,2}^2\right)^2 < \infty$ for $k_j \leq i < k_j + a$, where

$$\sigma_{j;1}^2 = \mathbb{E}\left[W_{j,i;1}^2\right], \text{ and } \sigma_{j;2}^2 = \mathbb{E}\left[W_{j,i;2}^2\right] \tag{57}$$

The three last displays imply (56) and thus that $B_{n_j}$ is bounded in probability. Moreover, since $f_{n_j}$ reads

$$\frac{1}{\sqrt{n_j}} \left| \sum_{i=1}^{\lfloor n_j\kappa\rfloor} \sigma_{j;1}^2 - \kappa \sum_{i=1}^{\lfloor n_j\kappa\rfloor} \sigma_{j;1}^2 - \kappa \sum_{i=\lfloor n_j\kappa\rfloor+1}^{\lfloor n_j\kappa\rfloor+a} \sigma_{j,i}^2 - \kappa \sum_{i=\lfloor n_j\kappa\rfloor+a+1}^{n_j} \sigma_{j;2}^2 \right|$$

$$= \sqrt{n_j}\kappa(1-\kappa)\left|\sigma_{j;1}^2 - \sigma_{j;2}^2\right| + O(n_j^{-1/2}),$$

we get

$$S_{n_j}(\kappa) - \kappa S_{n_j}(1) = \sqrt{n_j}\kappa(1-\kappa)\left(\sigma_{j;1}^2 - \sigma_{j;2}^2\right) + O_P(1). \tag{58}$$

We now study the denominator $s_{j,n_j}^2$ in (13). Denote by

$$\bar{\sigma}_j^2 = \frac{1}{n_j}\sum_{i=1}^{n_j} \sigma_{j,i}^2$$

the expectation of the scalogram (which now differs from the wavelet spectrum). Let us consider for $\tau \in \{0,\ldots,q(n_j)\}$ $\widehat{\gamma}_j(\tau)$ the empirical covariance of the wavelet coefficients defined in (15).

$$\widehat{\gamma}_j(\tau) = \frac{1}{n_j}\sum_{i=1}^{n_j-\tau}(W_{j,i}^2 - \bar{\sigma}_j^2)(W_{j,i+\tau}^2 - \bar{\sigma}_j^2) - \left(1 + \frac{\tau}{n_j}\right)\left(\bar{\sigma}_j^2 - \widehat{\sigma}_j^2\right)^2$$

$$+ \frac{1}{n_j}(\widehat{\sigma}_j^2 - \bar{\sigma}_j^2)\left\{\sum_{i=n_j-\tau+1}^{n_j}(W_{j,i}^2 - \bar{\sigma}_j^2) + \sum_{i=1}^{\tau}(W_{j,i}^2 - \bar{\sigma}_j^2)\right\}.$$

Using Minkowski inequality and (56), there exists a constant $C$ such that for all $1 \leq l \leq l + \tau \leq n_j$,

$$\left\| \sum_{i=l}^{l+\tau}(W_{j,i}^2 - \bar{\sigma}_j^2) \right\|_2 \leq \left\| \sum_{i=l}^{l+\tau}(W_{j,i}^2 - \sigma_{j,i}^2) \right\|_2 + \left\| \sum_{i=l}^{l+\tau}(\sigma_{j,i}^2 - \bar{\sigma}_j^2) \right\|_2$$

$$\leq C(\tau^{1/2} + \tau),$$

and similarly

$$\left\| \widehat{\sigma}_j^2 - \bar{\sigma}_j^2 \right\|_2 \le \frac{C}{\sqrt{n_j}}.$$

By combining these two latter bounds, the Cauchy-Schwarz inequality implies that

$$\left\| \frac{1}{n_j} \left( \widehat{\sigma}_j^2 - \bar{\sigma}_j^2 \right) \sum_{i=l}^{l+\tau} (W_{j,i}^2 - \bar{\sigma}_j^2) \right\|_1 \le \frac{C(\tau^{1/2} + \tau)}{n_j^{3/2}}.$$

Recall that $s_{j,n_j}^2 = \sum_{\tau=-q}^{q} w_\tau(q) \widehat{\gamma}_j(\tau)$ where $w_\tau(q)$ are the so-called Bartlett weights defined in (16). We now use the bounds above to identify the limit of $s_{j,n_j}^2$ as the sample size goes to infinity. The two previous identities imply that

$$\sum_{\tau=0}^{q} w_\tau(q) \left( 1 + \frac{\tau}{n_j} \right) \left\| \widehat{\sigma}_j^2 - \bar{\sigma}_j^2 \right\|_2 \le C \frac{q^2}{n_j^{3/2}}$$

and

$$\sum_{\tau=0}^{q} w_\tau(q) \left\| \frac{1}{n_j} \left( \widehat{\sigma}_j^2 - \bar{\sigma}_j^2 \right) \sum_{i=l}^{l+\tau} (W_{j,i}^2 - \bar{\sigma}_j^2) \right\|_1 \le C \frac{q^2}{n_j^{3/2}},$$

Therefore, we obtain

$$s_{j,n_j}^2 = \sum_{\tau=-q}^{q} w_\tau(q) \widehat{\gamma}_j(\tau) + O_P \left( \frac{q^2}{n_j^{3/2}} \right), \tag{59}$$

where $\widehat{\gamma}_j(\tau)$ is defined by

$$\widehat{\gamma}_j(\tau) = \frac{1}{n_j} \sum_{i=1}^{n_j - \tau} (W_{j,i}^2 - \bar{\sigma}_j^2)(W_{j,i+\tau}^2 - \bar{\sigma}_j^2). \tag{60}$$

Observe that since $q = o(n_j)$, $k_j = \lfloor n_j \kappa \rfloor$ and $0 \le \tau \le q$, then for any given integer $a$ and $n$ large enough $0 \le \tau \le k_j \le k_j + a \le n_j - \tau$ thus in (60) we may write $\sum_{i=1}^{n_j - \tau} = \sum_{i=1}^{k_j - \tau} + \sum_{i=k_j - \tau+1}^{k+a} + \sum_{i=k_j + a+1}^{n_j - \tau}$. Using $\sigma_{j;1}^2$ and $\sigma_{j;2}^2$ in (60) and straightforward bounds that essentially follow from (56), we get $s_{j,n_j}^2 = \tilde{s}_{j,n_j}^2 + O_P \left( \frac{q^2}{n_j} \right)$, where

$$\tilde{s}_{j,n_j}^2 = \sum_{\tau=-q}^{q} w_\tau(q) \left( \frac{k}{n_j} \widetilde{\gamma}_{j;1}(\tau) + \frac{n_j - k_j - a}{n_j} \widetilde{\gamma}_{j;2}(\tau) \right.$$
$$\left. + \frac{k_j - |\tau|}{n_j} \left( \sigma_{j;1}^2 - \bar{\sigma}_j^2 \right)^2 + \frac{n_j - k_j - a - |\tau|}{n_j} \left( \sigma_{j;2}^2 - \bar{\sigma}_j^2 \right)^2 \right)$$

with

$$\tilde{\gamma}_{j;1}(\tau) = \frac{1}{k_j} \sum_{i=1}^{k_j - \tau} \left( W_{j,i}^2 - \sigma_{j;1}^2 \right) \left( W_{j,i+\tau}^2 - \sigma_{j;1}^2 \right),$$

$$\tilde{\gamma}_{j;2}(\tau) = \frac{1}{n_j - k_j - a} \sum_{i=k_j+a+1}^{n_j - \tau} \left( W_{j,i}^2 - \sigma_{j;2}^2 \right) \left( W_{j,i+\tau}^2 - \sigma_{j;2}^2 \right).$$

Using that $\bar{\sigma}_j^2 \to \kappa \sigma_{j;1}^2 + (1 - \kappa) \sigma_{j;2}^2$ as $n_j \to \infty$, and that, for $\varepsilon = 1, 2$,

$$s_{j,n_j;\varepsilon}^2 \stackrel{\text{def}}{=} \sum_{\tau=-q}^{q} w_\tau(q) \tilde{\gamma}_{j;\varepsilon}(\tau) \stackrel{P}{\longrightarrow} \frac{1}{\pi} \int_{-\pi}^{\pi} |\mathbf{D}_{j,0;\varepsilon}(\lambda)|^2 d\lambda ,$$

we obtain

$$s_{j,n_j}^2 = \frac{1}{\pi} \int_{-\pi}^{\pi} \left\{ \kappa |\mathbf{D}_{j,0;1}(\lambda)|^2 + (1 - \kappa) |\mathbf{D}_{j,0;2}(\lambda)|^2 \right\} d\lambda$$

$$2q\kappa(1 - \kappa) \left( \sigma_{j;1}^2 - \sigma_{j;2}^2 \right)^2 + + o_p(1) + O_P\left( \frac{q^2}{n_j} \right). \quad (61)$$

Using (58), the last display and that $o_P(1) + O_P\left( \frac{q^2}{n_j} \right) = o_p(q)$, we finally obtain

$$S_{n_j}(\kappa) - \kappa S_{n_j}(1) = \frac{\sqrt{n_j} \kappa(1 - \kappa) \left| \sigma_{j;1}^2 - \sigma_{j;2}^2 \right| + O_P(1)}{\sqrt{2q(\kappa(1 - \kappa))} \left| \sigma_{j;1}^2 - \sigma_{j;2}^2 \right| + o_p(\sqrt{q})}$$

$$= \frac{\sqrt{n_j}}{\sqrt{2q}} \sqrt{\kappa(1 - \kappa)} (1 + o_p(1)) ,$$

which concludes the proof of Theorem 5.1.

### 5.2 Power of the test in multiple scales case

The results obtained in the previous Section in the single scale case easily extend to the test procedure designed to handle the multiple scales case. The alternative is specified exactly in the same way than in the single scale case but instead of considering the square of the wavelet coefficients at a given scale, we now study the behavior of the between-scale process. Consider the following process for $\varepsilon = 1, 2$,

$$Y_{J_1, J_2, i; \varepsilon} = \left( W_{J_2, i; \varepsilon}^2, \sum_{u=1}^{2} W_{J_2-1, 2(i-1)+u; \varepsilon}^2, \cdots, \sum_{u=1}^{2^{(J_2-J_1)}} W_{J_1, 2^{(J_2-J_1)}(i-1)+u; \varepsilon}^2 \right)^T ,$$

where $J_1$ and $J_2$ are respectively the finest and the coarsest scale considered in the test, $W_{j,i;\varepsilon}$ are defined in (53) and (54) and $\Gamma_{J_1,J_2;\varepsilon}$ the $(J_2 - J_1 + 1) \times (J_2 - J_1 + 1)$ symmetric non negative matrix such that

$$\Gamma_{J_1,J_2;\varepsilon}(j,j') = \sum_{h \in \mathbb{Z}} \mathrm{cov}(Y_{j,0;\varepsilon}, Y_{j',h;\varepsilon}) = \int_{-\pi}^{\pi} \left\| \mathbf{D}_{j,u;\varepsilon}(\lambda;f) \right\|^2 d\lambda, \qquad (62)$$

with $1 \le j, j' \le J_2 - J_1 + 1$ for $\varepsilon = 1,2$.

**Theorem 5.2.** *Consider $\{X_k^n\}_{k \in \mathbb{Z}}$ be a sequence of processes specified by (47) and (48). Finally assume that for at least one $j \in \{J_1, \dots, J_2\}$ and that at least one of the two matrices $\Gamma_{J_1,J_2;\varepsilon}$ $\varepsilon = 1,2$ defined in (62) is positive definite. Assume in addition that Finally, assume that the number of lags $q(n_{J_2})$ in the Bartlett estimate of the covariance matrix (36) is non decreasing and:*

$$q(n_{j_2}) \to \infty \quad and \quad \frac{q^2(n_{J_2})}{n_{J_2}} \to 0, \ as \ n_{J_2} \to \infty, \ . \qquad (63)$$

*Then, the W2-CUSUM test statistics $T_{J_1,J_2}$ defined by (41) satisfies*

$$\frac{n_{J_2}}{2q(n_{J_2})} \kappa(1 - \kappa)(1 + o_P(1)) \le T_{J_1,J_2} \xrightarrow{P} \infty \ as \quad n_{J_2} \to \infty$$

*Proof.* As in the single scale case we drop the dependence in $n_{J_2}$ in the expression of $q$ in this proof section. Let $k_j = \lfloor n_j \kappa \rfloor$ the change point in the wavelet spectrum at scale $j$. Then using (38) we have that $T_{J_1,J_2} \ge S_{J_1,J_2}(\kappa) - \kappa S_{J_1,J_2}(1)$ where

$$S_{J_1,J_2}(\kappa) - \kappa S_{J_1,J_2}(1) = \frac{1}{\sqrt{n_{J_2}}} \left[ n_j (B_{n_j} + f_{n_j}) \right]_{j=J_1,\dots,J_2},$$

where $B_{n_j}$ and $f_{n_j}$ are defined respectively by (51) and (52). Hence as in (58), we have

$$S_{J_1,J_2}(\kappa) - \kappa S_{J_1,J_2}(1) = \sqrt{n_{J_2}} \kappa(1 - \kappa)\Delta + O_P(1),$$

where $\Delta = \left[ \sigma^2_{J_1,J_2;1} - \sigma^2_{J_1,J_2;2} \right]^T$ and

$$\sigma^2_{J_1,J_2;\varepsilon} = \left( \sigma^2_{J_2;\varepsilon}, \dots, 2^{J_2 - J_1} \sigma^2_{J_1;\varepsilon} \right)^T.$$

We now study the asymptotic behavior of $\widehat{\Gamma}_{J_1,J_2}$. Using similar arguments as those leading to (61) in the proof of Theorem 5.1, we have

$$\widehat{\Gamma}_{J_1,J_2} = 2q\kappa(1 - \kappa)\Delta\Delta^T + \kappa\Gamma_{J_1,J_2;1} + (1 - \kappa)\Gamma_{J_1,J_2;2}$$

$$+ O_P\left( \frac{q}{n_{J_2}} \right) + O_P(q^{-1}) + O_P\left( \frac{q^2}{n_{J_2}} \right).$$

For $\Gamma$ a positive definite matrix, consider the matrix $\mathbf{M}(\Gamma) = \Gamma + 2q\kappa(1-\kappa)\Delta\Delta^T$. Using the matrix inversion lemma, the inverse of $\mathbf{M}(\Gamma)$ may be expressed as

$$\mathbf{M}^{-1}(\Gamma) = \left( \Gamma^{-1} - \frac{2q\kappa(1-\kappa)\Gamma^{-1}\Delta\Delta^T\Gamma^{-1}}{1+2q\kappa(1-\kappa)\Delta^T\Gamma^{-1}\Delta} \right),$$

which implies that

$$\Delta^T\mathbf{M}^{-1}(\Gamma)\Delta = \frac{\Delta^T\Gamma^{-1}\Delta}{1+2q\kappa(1-\kappa)\Delta^T\Gamma^{-1}\Delta}.$$

Applying these two last relations to $\Gamma_0 = \kappa\Gamma_{J_1,J_2}^{(1)} + (1-\kappa)\Gamma_{J_1,J_2}^{(2)}$ which is symmetric and definite positive (since, under the stated assumptions at least one of the two matrix $\Gamma_{J_1,J_2;\varepsilon}$, $\varepsilon = 1,2$ is positive) we have

$$T_{J_1,J_2} \geq \kappa^2(1-\kappa)^2 n_{J_2} \Delta^T \mathbf{M}^{-1}\left( \Gamma_0 + O_P\left(\frac{q^2}{n_{J_2}}\right) + O_P(q^{-1}) \right)\Delta + O_P(1)$$

$$= n_{J_2}\kappa^2(1-\kappa)^2 \frac{\Delta^T\Gamma_0^{-1}\Delta + O_P\left(\frac{q^2}{n_{J_2}}\right) + O_P(q^{-1}}{2q\kappa(1-\kappa)\Delta^T\Gamma_0^{-1}\Delta(1+o_p(1))} + O_P(1)$$

$$= \frac{n_{J_2}}{2q}\kappa(1-\kappa)(1+o_p(1)).$$

Thus $T_{J_1,J_2} \xrightarrow{P} \infty$ as $n_{J_2} \to \infty$, which completes the proof of Theorem 5.2.

*Remark 0.3.* The term corresponding to the "bias" term $\kappa\Gamma_{J_1,J_2;1} + (1-\kappa)\Gamma_{J_1,J_2;2}$ in the single case is $\frac{1}{\pi}\int_{-\pi}^{\pi}\{\kappa|\mathbf{D}_{j,0;1}(\lambda)|^2 + (1-\kappa)|\mathbf{D}_{j,0;2}(\lambda)|^2\}d\lambda = O(1)$, which can be neglected since the main term in $s_{j,n_j}^2$ is of order $q \to \infty$. In multiple scale case, the main term in $\widehat{\Gamma}_{J_1,J_2}$ is still of order $q$ but is no longer invertible (the rank of the leading term is equal to 1). A closer look is thus necessary and the term $\kappa\Gamma_{J_1,J_2;1} + (1-\kappa)\Gamma_{J_1,J_2;2}$ has to be taken into account. This is also explains why we need the more stringent condition (63) on the bandwidth size in the multiple scales case.

# 6 Some examples

In this section, we report the results of a limited Monte-Carlo experiment to assess the finite sample property of the test procedure. Recall that the test rejects the null if either CVM$(J_1,J_2)$ or KSM$(J_1,J_2)$, defined in (43) and (45) exceeds the $(1-\alpha)$-th quantile of the distributions $C(J_2-J_1+1)$ and $D(J_2-J_1+1)$, specified in (44) and (46). The quantiles are reported in Tables (1) and (2), and have been obtained by truncating the series expansion of the cumulative distribution function. To study the influence on the test procedure of the strength of the dependency, we consider different classes of Gaussian processes, including white noise, autoregressive moving average (ARMA) processes as well as fractionally integrated ARMA ($ARFIMA(p,d,q)$) processes which are known to be long range dependent. In all the simulations we set the lowest scale to $J_1 = 1$ and vary the coarsest scale $J_2 = J$. We used a wide range of values of sample size $n$, of the number of scales $J$ and of the parameters of the ARMA and FARIMA processes but, to conserve space, we present the results only for $n = 512, 1024, 2048, 4096, 8192, J = 3, 4, 5$ and four different models: an AR(1) process with parameter 0.9, a MA(1) process with parameter 0.9, and two ARFIMA(1,d,1) processes with memory parameter $d = 0.3$ and $d = 0.4$, and the same AR and MA coefficients, set to 0.9 and 0.1. In our simulations, we have used the Newey-West estimate of the bandwidth $q(n_j)$ for the covariance estimator (as implemented in the R-package *sandwich*).

### 6.0.1 Asymptotic level of *KSM* and *CVM*.

We investigate the finite-sample behavior of the test statistics $CVM(J_1,J_2)$ and $KSM(J_1,J_2)$ by computing the number of times that the null hypothesis is rejected in 1000 independent replications of each of these processes under $\mathcal{H}_0$ , when the asymptotic level is set to 0.05.

| White noise | | | | | |
|---|---|---|---|---|---|
| $n$ | 512 | 1024 | 2048 | 4096 | 8192 |
| $J = 3$ *KSM* | 0.02 | 0.01 | 0.03 | 0.02 | 0.02 |
| $J = 3$ *CVM* | 0.05 | 0.045 | 0.033 | 0.02 | 0.02 |
| $J = 4$ *KSM* | 0.047 | 0.04 | 0.04 | 0.02 | 0.02 |
| $J = 4$ *CVM* | 0.041 | 0.02 | 0.016 | 0.016 | 0.01 |
| $J = 5$ *KSM* | 0.09 | 0.031 | 0.02 | 0.025 | 0.02 |
| $J = 5$ *CVM* | 0.086 | 0.024 | 0.012 | 0.012 | 0.02 |

**Table 3** Empirical level of KSM − CVM for a white noise.

| MA(1)$[\theta = 0.9]$ | | | | | |
|---|---|---|---|---|---|
| $n$ | 512 | 1024 | 2048 | 4096 | 8192 |
| $J = 3$ *KSM* | 0.028 | 0.012 | 0.012 | 0.012 | 0.02 |
| $J = 3$ *CVM* | 0.029 | 0.02 | 0.016 | 0.016 | 0.01 |
| $J = 4$ *KSM* | 0.055 | 0.032 | 0.05 | 0.025 | 0.02 |
| $J = 4$ *CVM* | 0.05 | 0.05 | 0.03 | 0.02 | 0.02 |
| $J = 5$ *KSM* | 0.17 | 0.068 | 0.02 | 0.02 | 0.02 |
| $J = 5$ *CVM* | 0.13 | 0.052 | 0.026 | 0.021 | 0.02 |

**Table 4** Empirical level of KSM − CVM for a $MA(q)$ process.

| AR(1)$[\phi = 0.9]$ | | | | | |
|---|---|---|---|---|---|
| $n$ | 512 | 1024 | 2048 | 4096 | 8192 |
| $J = 3$ KSM | 0.083 | 0.073 | 0.072 | 0.051 | 0.04 |
| $J = 3$ CVM | 0.05 | 0.05 | 0.043 | 0.032 | 0.03 |
| $J = 4$ KSM | 0.26 | 0.134 | 0.1 | 0.082 | 0.073 |
| $J = 4$ CVM | 0.14 | 0.092 | 0.062 | 0.04 | 0.038 |
| $J = 5$ KSM | 0.547 | 0.314 | 0.254 | 0.22 | 0.11 |
| $J = 5$ CVM | 0.378 | 0.221 | 0.162 | 0.14 | 0.093 |

**Table 5** Empirical level of KSM − CVM for an $AR(1)$ process.

| ARFIMA(1,0.3,1)$[\phi = 0.9, \theta = 0.1]$ | | | | | |
|---|---|---|---|---|---|
| $n$ | 512 | 1024 | 2048 | 4096 | 8192 |
| $J = 3$ KSM | 0.068 | 0.047 | 0.024 | 0.021 | 0.02 |
| $J = 3$ CVM | 0.05 | 0.038 | 0.03 | 0.02 | 0.02 |
| $J = 4$ KSM | 0.45 | 0.42 | 0.31 | 0.172 | 0.098 |
| $J = 4$ CVM | 0.39 | 0.32 | 0.20 | 0.11 | 0.061 |
| $J = 5$ KSM | 0.57 | 0.42 | 0.349 | 0.229 | 0.2 |
| $J = 5$ CVM | 0.41 | 0.352 | 0.192 | 0.16 | 0.11 |

**Table 6** Empirical level of KSM − CVM for an $ARFIMA(1, 0.3, 1)$ process.

| ARFIMA(1,0.4,1)$[\phi = 0.9, \theta = 0.1]$ | | | | | |
|---|---|---|---|---|---|
| $n$ | 512 | 1024 | 2048 | 4096 | 8192 |
| $J = 3$ KSM | 0.11 | 0.063 | 0.058 | 0.044 | 0.031 |
| $J = 3$ CVM | 0.065 | 0.05 | 0.043 | 0.028 | 0.02 |
| $J = 4$ KSM | 0.512 | 0.322 | 0.26 | 0.2 | 0.18 |
| $J = 4$ CVM | 0.49 | 0.2 | 0.192 | 0.16 | 0.08 |
| $J = 5$ KSM | 0.7 | 0.514 | 0.4 | 0.321 | 0.214 |
| $J = 5$ CVM | 0.59 | 0.29 | 0.262 | 0.196 | 0.121 |

**Table 7** Empirical level of KSM − CVM for an $ARFIMA(1, 0.3, 1)$ process.

We notice that in general the empirical levels for the CVM are globally more accurate than the ones for the KSM test, the difference being more significant when the strength of the dependence is increased, or when the number of scales that are tested simultaneously get larger. The tests are slightly too conservative in the white noise and the MA case (tables (3) and (4)); in the AR(1) case and in the ARFIMA cases, the test rejects the null much too often when the number of scales is large compared to the sample size (the difficult problem being in that case to estimate the covariance matrix of the test). For $J = 4$, the number of samples required to meet the target rejection rate can be as large as $n = 4096$ for the CVM test and $n = 8192$ for the KSM test. The situation is even worse in the ARFIMA case (tables (6) and (7)). When the number of scales is equal to 4 or 5, the test rejects the null hypothesis much too often.

Pvalues under H0 of white noise–AR(1)–MA(1)

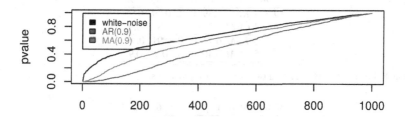

Pvalues under H0 of ARFIMA(0.9,d,0.1)

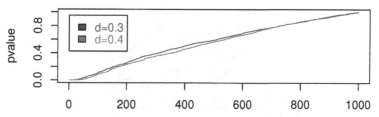

**Fig. 2** P-value under $\mathcal{H}_0$ of the distribution $D(J)$ $n = 1024$ for white noise and MA(1) processes and $n = 4096$ for AR(1) and ARFIMA(1,d,1) processes; the coarsest scale is $J = 4$ for white noise, MA and AR processes and $J = 3$ for the ARFIMA process. The finest scale is $J_1 = 1$.

### 6.0.2 Power of *KSM* and *CVM*.

We assess the power of test statistic by computing the test statistics in presence of a change in the spectral density. To do so, we consider an observation obtained by concatenation of $n_1$ observations from a first process and $n_2$ observations from a second process, independent from the first one and having a different spectral density. The length of the resulting observations is $n = n_1 + n_2$. In all cases, we set $n_1 = n_2 = n/2$, and we present the results for $n_1 = 512, 1024, 2048, 4096$ and scales $J = 4, 5$. We consider the following situations: the two processes are white Gaussian noise with two different variances, two AR processes with different values of the autoregressive coefficient, two MA processes with different values of the moving average coefficient and two ARFIMA with same moving average and same autoregressive coefficients but different values of the memory parameter $d$. The scenario considered is a bit artificial but is introduced here to assess the ability of the test to detect abrupt changes in the spectral content. For 1000 simulations, we report the number of times $\mathcal{H}_1$ was accepted, leading the following results.

The power of our two statistics gives us satisfying results for the considered processes, especially if the sample size tends to infinity.

| white-noise | $[\sigma_1^2 = 1, \sigma_2^2 = 0.7]$ | | | |
|---|---|---|---|---|
| $n_1 = n_2$ | 512 | 1024 | 2048 | 4096 |
| $J = 4$ KSM | 0.39 | 0.78 | 0.89 | 0.95 |
| $J = 4$ CVM | 0.32 | 0.79 | 0.85 | 0.9 |
| $J = 5$ KSM | 0.42 | 0.79 | 0.91 | 0.97 |
| $J = 5$ CVM | 0.40 | 0.78 | 0.9 | 0.9 |

**Table 8** Power of KSM − CVM on two white noise processes.

| MA(1)+MA(1) | | $[\theta_1 = 0.9, \theta_2 = 0.5]$ | | | |
|---|---|---|---|---|---|
| $n_1 = n_2$ | | 512 | 1024 | 2048 | 4096 |
| $J = 4$ | KSM | 0.39 | 0.69 | 0.86 | 0.91 |
| $J = 4$ | CVM | 0.31 | 0.6 | 0.76 | 0.93 |
| $J = 5$ | KSM | 0.57 | 0.74 | 0.84 | 0.94 |
| $J = 5$ | CVM | 0.46 | 0.69 | 0.79 | 0.96 |

**Table 9** Power of KSM − CVM on a concatenation of two different *MA* processes.

| AR(1)+AR(1) | | $[\phi_1 = 0.9, \phi_2 = 0.5]$ | | | |
|---|---|---|---|---|---|
| $n_1 = n_2$ | | 512 | 1024 | 2048 | 4096 |
| $J = 4$ | KSM | 0.59 | 0.72 | 0.81 | 0.87 |
| $J = 4$ | CVM | 0.53 | 0.68 | 0.79 | 0.9 |
| $J = 5$ | KSM | 0.75 | 0.81 | 0.94 | 0.92 |
| $J = 5$ | CVM | 0.7 | 0.75 | 0.89 | 0.91 |

**Table 10** Power of KSM − CVM on a concatenation of two different *AR* processes.

| ARFIMA(1,0.3,1) | + | ARFIMA(1,0.4,1) | $[\phi = 0.9, \theta = 0.1]$ | | |
|---|---|---|---|---|---|
| $n_1 = n_2$ | | 512 | 1024 | 2048 | 4096 |
| $J = 4$ | KSM | 0.86 | 0.84 | 0.8 | 0.81 |
| $J = 4$ | CVM | 0.81 | 0.76 | 0.78 | 0.76 |
| $J = 5$ | KSM | 0.94 | 0.94 | 0.9 | 0.92 |
| $J = 5$ | CVM | 0.93 | 0.92 | 0.96 | 0.91 |

**Table 11** Power of KSM − CVM two ARFIMA(1,d,1) with same AR and MA part but two different values of memory parameter $d$.

### 6.0.3 Estimation of the change point in the original process.

We know that for each scale $j$, the number $n_j$ of wavelet coefficients is $n_j = 2^{-j}(n - \Lambda + 1) - \Lambda + 1$. If we denote by $k_j$ the change point in the wavelet coefficients at scale $j$ and $k$ the change point in the original signal, then $k = 2^j(k_j + \Lambda - 1) + \Lambda - 1$. In this paragraph, we estimate the change point in the generalized spectral density of a process when it exists and give its 95% confidence interval. For that, we proceed as before. We consider an observation obtained by concatenation of $n_1$ observations from a first process and $n_2$ observations from a second process, independent from the first one and having a different spectral density. The length of the resulting observations is $n = n_1 + n_2$. we estimate the change point in the process and we present the result for $n_1 = 512, 1024, 4096, 8192, n_2 = 512, 2048, 8192, J =$

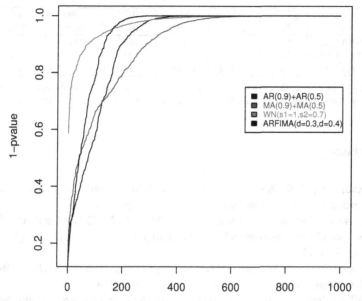

**Fig. 3** Empirical power of KSM($d = 4$) for white noise, AR, MA and ARFIMA processes.

3, the statistic *CVM*, two *AR* processes with different values of the autoregressive coefficient and two *ARFIMA* with same moving average and same autoregressive coefficients but different values of the memory parameter $d$. For 10000 simulations, the bootstrap confidence intervals obtained are set in the tables below. we give also the empirical mean and the median of the estimated change point.

- $[AR(1), \phi = 0.9]$ and $[AR(1), \phi = 0.5]$

| $n_1$ | 512 | 512 | 512 | 1024 | 4096 | 8192 |
|---|---|---|---|---|---|---|
| $n_2$ | 512 | 2048 | 8192 | 1024 | 4096 | 8192 |
| $MEAN_{CVM}$ | 478 | 822 | 1853 | 965 | 3945 | 8009 |
| $MEDIAN_{CVM}$ | 517 | 692 | 1453 | 1007 | 4039 | 8119 |
| $IC_{CVM}$ | [283,661] | [380,1369] | [523,3534] | [637,1350] | [3095,4614] | [7962,8825] |

**Table 12** Estimation of the change point and confidence interval at 95% in the generalized spectral density of a process which is obtain by concatenation of two AR(1) processes.

- $[ARFIMA(1, 0.2, 1)]$ and $[ARFIMA(1, 0.3, 1)]$, with $\phi = 0.9$ and $\theta = 0.2$

We remark that the change point belongs always to the considered confidence interval excepted for $n_1 = 512$, $n_2 = 8192$ where the confidence interval is $[523, 3534]$ and the change point $k = 512$ doesn't belong it. One can noticed that when the size of the sample increases and $n_1 = n_2$, the interval becomes more accurate. However, as expected, this interval becomes less accurate when the change appears either at the beginning or at the end of the observations.

| $n_1$ | 512 | 512 | 512 | 1024 | 4096 | 8192 |
|---|---|---|---|---|---|---|
| $n_2$ | 512 | 2048 | 8192 | 1024 | 4096 | 8192 |
| $MEAN_{CVM}$ | 531 | 1162 | 3172 | 1037 | 4129 | 8037 |
| $MEDIAN_{CVM}$ | 517 | 1115 | 3215 | 1035 | 4155 | 8159 |
| $IC_{CVM}$ | [227,835] | [375,1483] | [817,6300] | [527,1569] | [2985,5830] | [6162,9976] |

**Table 13** Estimation of the change point and confidence interval at 95% in the generalized spectral density of a process which is obtain by concatenation of two ARFIMA(1,d,1) processes.

# References

[1] Andrews., D. W. K. (1991) Heteroskedasticity and autocorrelation consistent covariance matrix estimation. *Econometrica*, **59**, 817–858.

[2] Berkes, I, Horvátz, L, Kokoszka P. and Shao, Q-M. (2006) On discriminating between long-range dependence and change in mean,. *The annals of statistics*, **34**, 1140–1165.

[3] Bhattacharya, R.N. Gupta, V.K. and Waymire, E. (1983) The Hurst effect under trends. *Journal of Applied Probability*, **20**, 649–662.

[4] Billingsley, P. (1999) *Convergence of probability measures*. Wiley Series in Probability and Statistics: Probability and Statistics. John Wiley & Sons Inc., New York, second edition.

[5] Boes, D.C. and Salas. J. D. (1978) Nonstationarity of the Mean and the Hurst Phenomenon. *Water Resources Research*, **14**, 135–143.

[6] Brodsky, B. E. and Darkhovsky, B. S. (2000) *Non-parametric statistical diagnosis*, volume 509 of *Mathematics and its Applications*. Kluwer Academic Publishers, Dordrecht, 2000.

[7] Carmona, P. Petit, F., Pitman, J. and Yor. M. (1999) On the laws of homogeneous functionals of the Brownian bridge. *Studia Sci. Math. Hungar.*, **35**, 445–455.

[8] Cohen, A. (2003) *Numerical analysis of wavelet methods*, volume 32 of *Studies in Mathematics and its Applications*. North-Holland Publishing Co., Amsterdam.

[9] Diebold, F.X. and Inoue, A. (2001) Long memory and regime switching. *J. Econometrics*, **105**, 131–159.

[10] Giraitis, L., Kokoszka, P., Leipus, R. and Teyssière, G. (2003) Rescaled variance and related tests for long memory in volatility and levels. *Journal of econometrics*, **112**, 265–294.

[11] Granger, C.W.J. and Hyung, N. (2004) Occasional structural breaks and long memory with an application to the S& P 500 absolute stock returns. *Journal of Empirical Finance*, **11**, 399–421.

[12] Hidalgo, J. and Robinson, P. M. (1996) Testing for structural change in a long-memory environment. *J. Econometrics*, **70**, 159–174.

[13] Hurvich, C.M., Lang, G. and Soulier, P. (2005) Estimation of long memory in the presence of a smooth nonparametric trend. *J. Amer. Statist. Assoc.*, **100**, 853–871.

[14] Inclan, C. and Tiao, G. C. (1994) Use of cumulative sums of squares for retrospective detection of changes of variance. *American Statistics*, **89**, 913–923.

[15] Kiefer, J. (1959) $K$-sample analogues of the Kolmogorov-Smirnov and Cramér-V. Mises tests. *Ann. Math. Statist.*, **30**, 420–447.

[16] Klemes, V. (1974) The Hurst Phenomenon: A Puzzle? *Water Resources Research*, **10** 675–688.

[17] Mallat, S. (1998) *A wavelet tour of signal processing*. Academic Press Inc., San Diego, CA.

[18] Mikosch, T. and Stăricŭ, C. (2004) Changes of structure in financial time series and the Garch model. *REVSTAT*, **2**, 41–73.

[19] Moulines, E., Roueff, F. and Taqqu, M. S. (2007) On the spectral density of the wavelet coefficients of long memory time series with application to the log-regression estimation of the memory parameter. *J. Time Ser. Anal.*, **28**, 155–187.

[20] Moulines, E., Roueff, F. and Taqqu M.S. (2008) A wavelet Whittle estimator of the memory parameter of a non-stationary Gaussian time series. *Ann. Statist.*, **36**, 1925–1956.

[21] Newey, W. K. and West K. D. (1987) A simple, positive semidefinite, heteroskedasticity and autocorrelation consistent covariance matrix. *Econometrica*, **55**, 703–708.

[22] Newey, W. K. and West K. D. (1994) Automatic lag selection in covariance matrix estimation. *Rev. Econom. Stud.*, **61**, 631–653.

[23] Pitman, J. and Yor, M. (1999). The law of the maximum of a Bessel bridge. *Electron. J. Probab.*, **4**, no. 15, 35 pp.

[24] Potter., K.W (1976) Evidence of nonstationarity as a physical explanation of the Hurst phenomenon. *Water Resources Research*, **12**, 1047–1052.

[25] Ramachandra Rao, A. and Yu., G.H. (1986) Detection of nonstationarity in hydrologic time series. *Management Science*, **32**, 1206–1217.

[26] Whitcher, B., Byers, S.D., Guttorp, P. and Percival, D. (2002) Testing for homogeneity of variance in time series : Long memory, wavelets and the nile river. *Water Resour. Res*, **38** (5), 1054, doi:10.1029/2001WR000509.

# Lecture Notes in Statistics

For information about Volumes 1 to 142, please contact Springer-Verlag

143: Russell Barton, Graphical Methods for the Design of Experiments. x, 208 pp., 1999.

144: L. Mark Berliner, Douglas Nychka, and Timothy Hoar (Editors), Case Studies in Statistics and the Atmospheric Sciences. x, 208 pp., 2000.

145: James H. Matis and Thomas R. Kiffe, Stochastic Population Models. viii, 220 pp., 2000.

146: Wim Schoutens, Stochastic Processes and Orthogonal Polynomials. xiv, 163 pp., 2000.

147: Jürgen Franke, Wolfgang Härdle, and Gerhard Stahl, Measuring Risk in Complex Stochastic Systems. xvi, 272 pp., 2000.

148: S.E. Ahmed and Nancy Reid, Empirical Bayes and Likelihood Inference. x, 200 pp., 2000.

149: D. Bosq, Linear Processes in Function Spaces: Theory and Applications. xv, 296 pp., 2000.

150: Tadeusz Caliński and Sanpei Kageyama, Block Designs: A Randomization Approach, Volume I: Analysis. ix, 313 pp., 2000.

151: Håkan Andersson and Tom Britton, Stochastic Epidemic Models and Their Statistical Analysis. ix, 152 pp., 2000.

152: David Ríos Insua and Fabrizio Ruggeri, Robust Bayesian Analysis. xiii, 435 pp., 2000.

153: Parimal Mukhopadhyay, Topics in Survey Sampling. x, 303 pp., 2000.

154: Regina Kaiser and Agustín Maravall, Measuring Business Cycles in Economic Time Series. vi, 190 pp., 2000.

155: Leon Willenborg and Ton de Waal, Elements of Statistical Disclosure Control. xvii, 289 pp., 2000.

156: Gordon Willmot and X. Sheldon Lin, Lundberg Approximations for Compound Distributions with Insurance Applications. xi, 272 pp., 2000.

157: Anne Boomsma, Marijtje A.J. van Duijn, and Tom A.B. Snijders (Editors), Essays on Item Response Theory. xv, 448 pp., 2000.

158: Dominique Ladiray and Benoît Quenneville, Seasonal Adjustment with the X-11 Method. xxii, 220 pp., 2001.

159: Marc Moore (Editor), Spatial Statistics: Methodological Aspects and Some Applications. xvi, 282 pp., 2001.

160: Tomasz Rychlik, Projecting Statistical Functionals. viii, 184 pp., 2001.

161: Maarten Jansen, Noise Reduction by Wavelet Thresholding. xxii, 224 pp., 2001.

162: Constantine Gatsonis, Bradley Carlin, Alicia Carriquiry, Andrew Gelman, Robert E. Kass Isabella Verdinelli, and Mike West (Editors), Case Studies in Bayesian Statistics, Volume V. xiv, 448 pp., 2001.

163: Erkki P. Liski, Nripes K. Mandal, Kirti R. Shah, and Bikas K. Sinha, Topics in Optimal Design. xii, 164 pp., 2002.

164: Peter Goos, The Optimal Design of Blocked and Split-Plot Experiments. xiv, 244 pp., 2002.

165: Karl Mosler, Multivariate Dispersion, Central Regions and Depth: The Lift Zonoid Approach. xii, 280 pp., 2002.

166: Hira L. Koul, Weighted Empirical Processes in Dynamic Nonlinear Models, Second Edition. xiii, 425 pp., 2002.

167: Constantine Gatsonis, Alicia Carriquiry, Andrew Gelman, David Higdon, Robert E. Kass, Donna Pauler, and Isabella Verdinelli (Editors), Case Studies in Bayesian Statistics, Volume VI. xiv, 376 pp., 2002.

168: Susanne Rässler, Statistical Matching: A Frequentist Theory, Practical Applications and Alternative Bayesian Approaches. xviii, 238 pp., 2002.

169: Yu. I. Ingster and Irina A. Suslina, Nonparametric Goodness-of-Fit Testing Under Gaussian Models. xiv, 453 pp., 2003.

170: Tadeusz Caliński and Sanpei Kageyama, Block Designs: A Randomization Approach, Volume II: Design. xii, 351 pp., 2003.

171: D.D. Denison, M.H. Hansen, C.C. Holmes, B. Mallick, B. Yu (Editors), Nonlinear Estimation and Classification. x, 474 pp., 2002.

172: Sneh Gulati, William J. Padgett, Parametric and Nonparametric Inference from Record-Breaking Data. ix, 112 pp., 2002.

173: Jesper Møller (Editor), Spatial Statistics and Computational Methods. xi, 214 pp., 2002.

174: Yasuko Chikuse, Statistics on Special Manifolds. xi, 418 pp., 2002.

175: Jürgen Gross, Linear Regression. xiv, 394 pp., 2003.

176: Zehua Chen, Zhidong Bai, Bimal K. Sinha, Ranked Set Sampling: Theory and Applications. xii, 224 pp., 2003.

177: Caitlin Buck and Andrew Millard (Editors), Tools for Constructing Chronologies: Crossing Disciplinary Boundaries, xvi, 263 pp., 2004.

178: Gauri Sankar Datta and Rahul Mukerjee , Probability Matching Priors: Higher Order Asymptotics, x, 144 pp., 2004.

179: D.Y. Lin and P.J. Heagerty (Editors), Proceedings of the Second Seattle Symposium in Biostatistics: Analysis of Correlated Data, vii, 336 pp., 2004.

180: Yanhong Wu, Inference for Change-Point and Post-Change Means After a CUSUM Test, xiv, 176 pp., 2004.

181: Daniel Straumann, Estimation in Conditionally Heteroscedastic Time Series Models , x, 250 pp., 2004.

182: Lixing Zhu, Nonparametric Monte Carlo Tests and Their Applications, xi, 192 pp., 2005.

183: Michel Bilodeau, Fernand Meyer, and Michel Schmitt (Editors), Space, Structure and Randomness, xiv, 416 pp., 2005.

184: Viatcheslav B. Melas, Functional Approach to Optimal Experimental Design, vii., 352 pp., 2005.

185: Adrian Baddeley, Pablo Gregori, Jorge Mateu, Radu Stoica, and Dietrich Stoyan, (Editors), Case Studies in Spatial Point Process Modeling, xiii., 324 pp., 2005.

186: Estela Bee Dagum and Pierre A. Cholette, Benchmarking, Temporal Distribution, and Reconciliation Methods for Time Series, xiv., 410 pp., 2006.

187: Patrice Bertail, Paul Doukhan and Philippe Soulier, (Editors), Dependence in Probability and Statistics, viii., 504 pp., 2006.

188: Constance van Eeden, Restricted Parameter Space Estimation Problems, vi, 176 pp., 2006.

189: Bill Thompson, The Nature of Statistical Evidence, vi, 152 pp., 2007.

190: Jérôme Dedecker, Paul Doukhan, Gabriel Lang, José R. León, Sana Louhichi Clémentine Prieur, Weak Dependence: With Examples and Applications, xvi, 336 pp., 2007.

191: Vlad Stefan Barbu and Nikolaos Limnios, Semi-Markov Chains and Hidden Semi-Markov Models toward Applications, xii, 228 pp., 2007.

192: David B. Dunson, Random Effects and Latent Variable Model Selection, 2008

193: Alexander Meister. Deconvolution Problems in Nonparametric Statistics, 2008.

194: Dario Basso, Fortunato Pesarin, Luigi Salmaso, Aldo Solari, Permutation Tests for Stochastic Ordering and ANOVA: Theory and Applications with R, 2009.

195: Alan Genz and Frank Bretz, Computation of Multivariate Normal and $t$ Probabilities, viii, 126 pp., 2009.

196: Hrishikesh D. Vinod, Advances in Social Science Research Using R, xx, 207 pp., 2010.

197: M. González, I.M. del Puerto, T. Martinez, M. Molina, M. Mota, A. Ramos (Eds.), Workshop on Branching Processes and Their Applications, xix, 296 pp., 2010.

198: Jaworski P, Durante F, Härdle W, Rychlik T. (Eds.), Copula Theory and Its Applications Proceedings of the Workshop Held in Warsaw, 25-26 September 2009 Subseries: Lecture Notes in Statistics - Proceedings, 327 pp., 50 illus., 25 in color., 2010.

199: Oja Hannu, Multivariate Nonparametric Methods with R An approach based on spatial signs and ranks Series: Lecture Notes in Statistics, xiii, 232 pp., 2010.

200: Doukhan Paul, Lang Gabriel, Teyssière Gilles (Eds.), Dependence in Probability and Statistics Series: Lecture Notes in Statistics, xv, 205 pp., 2010.